Modern physics
and antiphysics

adolph baker
lowell technological institute

addison-wesley publishing company
reading, massachusetts · menlo park, california
london · amsterdam · don mills, ontario · sydney

This book is in the
Addison-Wesley Series in Physics
David Lazarus, *Consulting Editor*

Second printing, May 1972

ISBN 0-201-00485-2
NOPQRSTUVW-AL-8987654321

*To the Lindas and Joels and Dannys and Ellies
who will soon inherit the earth in any case, one way
or the other.*

Foreword

This is an attempt to communicate some of the findings of twentieth century physics to people who have remained isolated from these developments. In order to accomplish this a style of presentation has been employed which is conventionally reserved for humanistic rather than scientific literature. The actual developments of modern physics have been reported as accurately and objectively as possible, but comments and interpretations of a social or philosophical nature are not purported to represent any monolithic position of the physics community, since none exists. An effort has been made to raise some of the relevant issues while exploring the physics, but certainly not to resolve them all. Many of the thoughts expressed in the dialogues are somewhat simplified distillations of ideas which are deliberately controversial or thought-provoking. Clearly they cannot all simultaneously reflect accurately the opinions of the author. If the protagonists appear at times overly defensive or dogmatic, this is not unlike the manner of most human beings engaged in similar arguments.

The physics questions appearing at the ends of chapters vary in difficulty, and some answers may be found in the back of the book. On the other hand, there are also questions which have social implications and do not have unique solutions; in such cases the reader is invited to answer according to his own outlook but in the context of the discussion. An Appendix has been provided for anyone who may wish to pursue the physical developments on a more mathematical level.

acknowledgments

For reading the manuscript in various stages of completion from the physicist's point of view, and offering comments and encouragement, I am grateful to David Lazarus, Kenneth W. Ford, Philip Morrison, Hugh Young, Robert H. March, William C. Rife, Marvin Lubin.

A number of "humanists" volunteered to represent and protect the interests of future readers. They include Joel and Linda Janowitz, Ellen Winner, Jack Karan, Joyce Rey, Lawrence Jaraslow, Nathan Pearlman.

Dora Baker was a source of ideas, sounding board, merciless critic, and boundless encourager. She never wavered in the conviction that this book ought to be written, and there is much of her in it.

Joyce Rey not only typed but listened to what she was typing and sounded an alarm when the reader was forgotten in the physics.

The selection by Edna St. Vincent Millay is from *Collected Poems,* Harper & Row, copyright 1934, 1962 by Edna St. Vincent Millay and Norma Millay Ellis. By permission of Norma Millay Ellis.

The couplet *Forgive, O Lord* is from *In the Clearing* by Robert Frost, copyright © 1962 by Robert Frost, reprinted by permission of Holt, Rinehart and Winston, Inc.

The quotation from Werner Heisenberg's *Physics and Philosophy,* copyright by him in 1958, is by permission of Harper & Row, publishers.

The quotation from Jean-Paul Sartre, *Existentialism and Human Emotions,* 1957, is by permission of Philosophical Library, publishers.

The photograph of Fraunhofer diffraction of light by a double slit is from Bruno Rossi, *Optics,* Addison Wesley Publishing Co., 1957.

Photographs of light and electron diffraction are from *Physics,* Physical Science Study Committee, 1965, and were supplied by the publisher, D.C. Heath & Co. Reproduced by permission of the original sources: Diffraction of light waves from Valasek, *Introduction to Theoretical and Experimental Optics*, John Wiley & Sons, 1949; the double slit interference of electrons from *Zeitschrift für Naturforschung;* and edge diffraction of electrons from *Handbuch der Physik*, Springer-Verlag.

The drawings are by Adolph Baker.

Wayland, Massachusetts   A.B.
January 1970

Contents

Prologue: The antiphysicists

O Earth, unhappy planet born to die,
Might I your scribe and your confessor be,
What wonders must you not relate to me
Of Man, who when his destiny was high
Strode like the sun into the middle sky
And shone an hour, and who so bright as he,
And like the sun went down into the sea,
Leaving no spark to be remembered by.

EDNA ST. VINCENT MILLAY
Epitaph for the Race of Man, Sonnet IV

No one knows when Linda started hating physics. It may have been the interminable dinner conversations, involving mother and father and a little brother who always blurted out the answers to the questions—animated discussions about angular momentum and limits and invariance principles, to which she remained a silent witness. It may have been a mathematics teacher who remembered his subject was a discipline but forgot that it was also a freedom. Or it may just have been a matter of temperament. It really does not matter. Linda belongs to a whole generation caught in the antiscience backlash. Born in the shadow of the atom bomb, raised in an economy of affluence, educated to the calls of the first Sputnik, and ordered to participate in a war which appeared to them to be basically immoral, an important section of this population has rejected both the Establishment and Science, its handmaiden.

This is the age of Marshall McLuhan, and television, and hot and cool media, and drugs, and campus sit-ins. The spirit of alienation sweeping the leading universities has been removing from academic life many of the very people who would have been the best scholars and future leaders, and at this writing no one knows how it will end.

It is possible that we shall come out ahead after all, that we were heading in the wrong direction, and that all this is a necessary precursor to a healthier society. But sooner or later the rebuilding must come, and it is hoped that such books as this can help to bridge the gap. It represents a peace offering. We begin by admitting that Science is not God, perhaps is not even necessary. Then we shall see if it is not at least interesting, perhaps even important.

Physics literature is most often directed toward the professional. When he tries to address himself to students, or to laymen, or to children, the physicist seems to have great difficulty in sloughing off the habit of writing over everyone's head. The idea is to have one's peers read the book, nod their heads, and say, "Hmm. Very good. Very neat." I should like to avoid this psychological trap by trying to exorcise the spectre of the physicist always looking over my shoulder. This book is not written for physicists, and it should therefore lack the elegance, the richness of content, and the conciseness of expression which are badges of the profession. If the reader and I are very lucky, the hope is that as a result it may succeed also in avoiding some of the obscurity. The spirit of such an undertaking could be characterized by a remark I once heard an excellent physicist make in introducing a relatively new subject, "I shall oversimplify, omit, perhaps even lie a little."

There is a large communication barrier between physicists and people. In an effort to breach this barrier a course in the elements of Physical Science is often required of that significant body of liberal arts students who, although intelligent, well-read, and open-minded, react negatively to scientific rigor and perhaps even traumatically to any suggestion of mathematical formalism. These courses have often been dismal failures. There is no such thing as instant physics. Somehow the sciences have failed to provide a favorable climate for amateurs and spectators. Yet it is possible to enjoy music without being able to play an instrument, or art without knowing how to paint. Millions of people (even some physicists) turn on television sets without knowledge or curiosity about the circuits which generate the picture or the sound.

I believe that if the average person had any inkling of what physicists have uncovered during the twentieth century, he would think himself to be a character in a science fiction novel. The future is now, and it is all buried in equations. I am going to try to convey some of the picture and the sound without getting very far into the hardware. Admittedly it is not as exciting to the reader as playing the game himself. But we all have to be spectators at something.

I have one request to make. This excursion may demand an adjustment in style of reading. Most humanistic literature has the property that ideas tend to be diffused over many pages. Thoughts may be introduced concurrently, but the immersion is gradual. The experienced individual develops the habit of rapid reading; in fact, comprehension may even improve with speed. The situation is quite different in the literature of physics and mathematics. There are no rewards for speed reading. The ideas come in the form of highly concentrated abstractions, organized in what McLuhan calls the "linear" structure, and all links are important. If the reader skims, he is likely to find himself suddenly in a strange land, with no means for communicating with the natives. Alienation sets in, and the journey is terminated. If this is to be avoided, we must both be prepared to compromise. I shall try wherever possible to make some adjustment, but the reader must also be willing to slow down a little. He may even find it useful to reread a passage from time to time. We shall have much shifting of gears as we pass back and forth between physics and other subjects, but with a little patience perhaps we can maintain our contact.

This book is written for nearly everyone. I shall be most happy if the antiphysicists will read it just for entertainment. And if some of them go so far as to acquire an interest in the developments of physics as a result, this will be an added dividend.

1   Is physics relevant?

*I do not know what I may appear to the world
but to myself I seem to have been only like a boy
playing on the seashore, and diverting myself
in now and then finding a smoother pebble
or a prettier shell than ordinary,
whilst the great ocean of truth
lay all undiscovered before me.*

SIR ISAAC NEWTON
*Brewster's Memoirs of Newton, Vol. II,
Chap. XXVII*

I heard a psychology professor give a lecture on the use of drugs, in which he said that marijuana is already being used by the majority of students, and that LSD is likely to revolutionize man's knowledge and understanding if only the government will permit experimentation. I asked him afterwards what he had observed to be the effect of drugs on the ability to function academically, and in particular whether there was not a correlation between the use of drugs and the increasingly prevalent conclusion being reached by students that studies are irrelevant. He leaned over and whispered, "Has it ever occurred to you that they might be right?"

It is very difficult to make an absolute value judgment of the relevancy of anything. I shall begin by making the assumption that there exists an objective universe which includes the people in it, and that it is interesting if not actually useful to know about this universe. If one accepts this, then one should not ask whether physics is relevant—not if one knows what physics is about. Of course there's the rub; most people do not know. My daughter Ellie came home from school one day and said,

"We learned in Science today what Physics is."

"And what is it?" I asked.

"Physics is the study of Dead Things," she said. "And Biology is the study of Living Things."

A most prevalent misconception about science is its confusion with technology. This is especially true in the United States, a highly pragmatic nation, which takes great pride in its plumbing and its can openers. No one used to pay much attention to science in this country, and scientists were traditionally regarded as eccentric cranks, until it was discovered that you can use science to blow up people. Then suddenly it became respectable, scientists became the darlings of the nation, and even presidents started asking their advice. Of course whenever people said science they really meant technology, but the scientists did not take great pains to correct this mistake. The most lavish spending has been in engineering applications, but physics has basked in the reflected glory of "applied science and development." Whenever a politician happened to read the fine print and asked, "But what is it for?" there was a ready answer: the science of today is the technology of tomorrow; and, one hastens to add, the weapons of the day after tomorrow.

Thus many young people are today being "turned off" by what they believe to be the scientific image. Even the government has become disenchanted. The weapons market appears to be saturated in an age of overkill, with many equivalent tons of TNT stockpiled for every living human being on this planet. And "science" did not win the war in Vietnam.

Let us clear up some misconceptions. It is not true, as often maintained, that the "purpose" of science is to serve man. It would be equally misleading, and in fact dangerous, to claim that the "purpose" of truth is to serve man; for in that case there are some among us who would hasten to suppress those truths which do not serve us well. Carried to its logical conclusion, this is the argument traditionally invoked by those defenders of the faith who believe that their unique historical role gives them the right to manipulate art or science or information in what they consider to be the public interest. Of course in its original form this propensity of human beings for teleological reasoning is merely an extension of the naiveté of children, who ask, "What is the rain for?" and are answered, "The rain is to cool us off," or, "To water the lawn," or, "To make puddles." Voltaire captured the quality of this egocentricism when he had Pangloss observe that noses were obviously made for glasses in this best of all possible worlds.

Science is not television, or tactical nuclear weapons, or rockets to the moon, or automatic Coke dispensers. Nor is it for that matter (and this may come as a shock) a cure for cancer. It is probably true that a genuine understanding of the life process would ultimately lead to a cure for cancer. It could also lead to a more efficient method for mass destruction. It is possible that when we eventually find a cure for cancer it will be because we have exhausted most of the possibilities. We may even stumble into it without understanding what it is—just as the drug Orinase was tried out for tuberculosis, and was found to control the patients' diabetes instead. I am not belittling medical research directed to finding cures for the diseases of mankind. There is no more urgent or worthwhile undertaking, and I often wish I were a part of it. But let us understand the difference between highly directed investigations on the one hand and basic research on the other.

The launching of Sputnik and the subsequent space achievements are regarded as great victories for physics. And they are. But the major developments of physics which made them possible took place hundreds of years ago. The Sputnik was a great victory for Isaac Newton. It also represents a magnificent achievement of modern engineering. It is important to note the distinction. When during World War II the physicists decided that they had to build nuclear weapons, they ceased for a time to be physicists and became nuclear engineers instead, even though this meant starting a new branch of engineering.

One may choose to quibble with this definition of science as simply the search for knowledge. It is not important, as long as the reader understands that in the context of this book science is defined as knowledge and the means for obtaining knowledge, without regard for the consequences. There is no morality in truth or knowledge. It is the use to which it is put which establishes morality.

The nature of scientific method is such that one must suppress one's hopes and wishes, and at some stages even one's intuition. In fact the distrust of self takes the form of setting traps to expose one's own fallacies. Only when a successful solution has been found can one be permitted the luxury of deciding whether the result is pleasant or useful. The student of physics has his intuition violated so repeatedly that he comes to accept it as a routine experience. When quantum mechanics was first developed in the 1920's in order to explain what had been observed in the laboratory, the implications were extremely painful to the physicists. What had come to be (and are still thought by most people to be) basic principles of scientific philosophy had to be reluctantly abandoned. Einstein, who had helped lead his people to the promised land, spent the remainder of his life looking for an alternative.

Most laymen, when they contemplate the effect physics may have had upon their lives, think of technology, war, automation. What they usually do not consider is the effect of science upon their way of reasoning. Psychiatrists interpret much of the instability in the world today as a product of the destruction of man's myths, which have always been a source of security. Among these were the myth of absolute truth and absolute right, the myth of determinism and predictability, and particularly the myth of the infallibility of established authority, including finally the authority of science itself.

It is customary to blame our sociological problems on the technological fallout resulting from the scientific revolution, but the really villainous act of science was the destruction of these myths. Furthermore, this time there are not even any new myths to replace the old ones. Man has recently discovered that the universe is not the beautifully structured machine his father and grandfather thought they lived in, and he is still reeling from the blow.

The great scientific and philosophical upsets in the first half of this century, as embodied in the developments of relativity and quantum mechanics, have played a significant role in creating the world in which today we find ourselves living and thinking. There ought therefore to be little doubt about the relevance not only of the findings of modern physics, but also of some of the techniques of reasoning which produced them. What is less obvious, but even more interesting, is the extent to which such techniques may be applied in coping with the problems of the world which they created. In the final analysis such persons as the reader will have a role to play in resolving this question. But he will not find the answer in a quick and superficial summary of the physics involved. We shall therefore explore these developments in some depth, subject of course to the restriction that the reader may not be versed in the mathematical techniques available to the physicists themselves. In accordance with the current trend toward participation, he will be given, if not equal time, at least a voice in the proceedings.

Poet You have not yet told me what physics is, except to suggest that it is not the study of dead things. Could it perhaps be a method for reducing everything to dead things?

Scientist You're still blaming me for the atom bomb.

P Not I. It's the voice of your conscience.

S In the first place you are confusing science with technology. And in the second place there were very good reasons for what we did during the Second World War. But in any case that is not what physics is about.

P It seems to me you run with the hare and hunt with the hounds. All right. Tell me what physics is about.

S The physicist looks for a structure which will enable him to make an orderly model of a chaotic universe. In principle it should eventually become possible to extend the methods of physics to all fields of knowledge, but in practice this still appears to be somewhat ambitious. For one thing there are difficulties with the mathematics in handling complex systems. For another some important links are still missing. But the objective is to achieve actual understanding, not just a collection of facts. To accomplish this physicists have to go wherever their travels take them. It is impossible to predict all the social implications of fundamental research. Columbus did not anticipate that he would discover a continent. He set out to test a theory, and to find a new path to the Orient. The important thing is that he went.

P But we already see where your travels are taking us. Your beautiful plastic-and-steel civilization has poisoned the air, turned cities into ghettos, and made mass murder our major national product. Soon perhaps the world will be a silent sepulchre, and there will be no one to hear the clicking of Geiger counters.

S   I do not share your pessimism. And I am making an effort to keep my temper. If we are to start placing blame, then tell me what contributions to our brave new civilization you have made lately. Even when nations are ruled by buffoons, the speeches are written by poets. Or is it perhaps the cigarette commercials you take pride in?

P   I do not write commercials.

S   And I do not make bombs.

2   The downfall of parity conservation, or how nature came to know right from left

Tiger, tiger burning bright
In the forests of the night,
What immortal hand or eye
Could frame thy fearful symmetry?

WILLIAM BLAKE
The Tiger

In a conventional book on modern physics, this subject would normally appear near the end. What I am about to describe represents one of the latest in the series of upsets and strange discoveries which have descended upon the scientific world since the turn of the century. Physicists love to arrange everything to be neat and tidy, and it is the custom to relate events chronologically, since this is how they occur. But we will forego this convention. We begin, not with the cataclysmic discoveries of the first half of this century, which have had such unmistakable effect on modern philosophical thought, but examine instead some recent developments which rather illustrate the elemental nature of the "simple" questions physicists try to answer.

When an artist looks at a landscape, no matter how rugged the terrain, his eye sees a measure of order and symmetry in what lies before him. Was the order always there, or has the artist introduced it? A physicist looks for symmetries in nature, and when he finds them, people are amazed at the extent of order which exists in nature. We walk around the universe, and, like Winnie the Pooh and Piglet, whenever we encounter our own footprints we say, "Aha! Someone has been here."

But even if a person is not an artist or a physicist, his instinctive response to a pattern or design is a quick search for possible symmetry. When he finds it, there is so much less for his eye to scan or his mind to remember. If the left half of a picture is like the right half, the field is immediately cut in two. If the top is also the same as the bottom, the region to be studied carefully is reduced by four. A child will more easily learn to recognize and reproduce the letter O than the letter C, because of its symmetry.

Physicists, like all human beings, are forced to cope with an apparently unreasonable and disorderly universe. The first step in solving a problem in physics is to look for symmetry, and thus reduce the difficulty. If the problem has symmetry, then so does the solution; thus even before we have done any work, we already know something about the answer.

Social problems are far more complex and therefore more difficult to solve than physics problems. But experience gained in simple problems often helps us in attacking the difficult ones. Physicists and mathematicians therefore always try to answer the easiest questions first. If the real world lacks symmetry, they may invent another world which has it, and begin by solving the problems in this more congenial world. The insight gained from the experience is then applied to the real problem. It is the same procedure followed by educators who introduce children to simple concepts before exposing them to all the complexities of life.

When genuine symmetries are found in nature, this is an enormous aid to understanding. Our knowledge of atomic and nuclear structure has in large measure been obtained by taking advantage of nature's symmetries. But recently the scientific world was amazed to discover that what had always been thought to be a universal symmetry was in fact no such thing at all.

What is symmetry? The word means a great deal more in the context of physics than it does in general. An artist might say that a design has symmetry if the left side "looks like" the right side. Let us put it more precisely. When a "symmetric" girl stands before a mirror, there should be no way for an observer to distinguish her from her mirror image; on the other hand, if she has a birthmark on her left cheek, then the "girl" in the mirror will instead have her birthmark on the right cheek. We should then say that our girl is unsymmetric, since we can always distinguish her from her mirror image. But in the case of a truly symmetric girl the birthmark either appears on both cheeks or not at all, and we cannot tell the real girl from the mirror girl.

The physicist regards this as only one special type of symmetry. What the mirror does is called a transformation of right into left, and vice versa. Let us now consider carefully what happens to an object or a physical process when it undergoes such a transformation, as it does when seen in a mirror. A right hand, for example, is transformed into a left hand. The left hand we see in the mirror is not a *real* left hand; it is merely the *image* of a right hand. But we do not have to look very far before we discover that this mirror image of a right hand corresponds to something which actually exists in nature, namely, a left hand. The very fact that such an image can have a counterpart in nature is considered by physicists to be a kind of symmetry, although it is somewhat more general and abstract than that of the girl in the mirror.

This extended definition may be looked upon as a "symmetry" between what is on *this side* of the mirror and what is on the *other side*. We know that what is on this side corresponds to reality, because it *is* reality. But we wonder about the quality of reality of what is on the other side. Could what we see in the mirror just as well be a real world? If so, we say this is a "symmetric" situation; if no real world could ever look or act like what we see in the mirror, we say the symmetry is "broken." If this appears a bit confusing, it is because the reader is still unaccustomed to having old words used for new ideas.

Suppose we now proceed to do all sorts of tricks in front of the mirror—make faces, turn cartwheels, hold political rallies, set off explosions. Always the image in the mirror turns out to be one which could also have occurred in nature, at least in principle. If an outside observer looks into the mirror and sees all these antics, he will have no way of knowing whether he is looking at real people or mirror people. This is due to the equality, or *parity*, which exists between right and left. The fact that a girl has a birthmark on her left cheek in no way precludes the existence somewhere in the universe of another girl with an identical birthmark on her right cheek. Nature has no reason to prefer one over the other. We call this lack of preference a symmetry. The expression used in the language of physics is that "parity is conserved" in all the processes observed in the course of our antics in front of the mirror. The term *parity* is reserved for this type of right-left symmetry, and, as we have seen, it is somewhat more abstract than that of an artist. When we say that parity is conserved, we mean that there is no way of distinguishing the mirror world from the real world.

Poet If I write a poem in front of a mirror, the poet I see in the mirror will use mirror writing, which would not look like any poem I have ever read. I can certainly distinguish this mirror poetry from real poetry, and it is immediately apparent that what I see could only happen in a mirror. Is not this a violation of your definition of "symmetry"? It seems to me that when it comes to printed words, "parity is *not* conserved," to use your language, despite the fact that it appears to be when you simply jump up and down in front of the mirror.

Scientist Mirror writing is not unnatural. It is only uncommon. One could write a poem which will look just like the one in the mirror. Perhaps somewhere in the universe that is how poetry is written.

Thou in thy lake dost see	ɘɘƨ ƚƨob ɘʞɒl ʏʜƚ ni uoʜT
Thyself: so she	ɘʜƨ oƨ :ʇlɘƨʏʜT
Beholds her image in her eyes	ƨɘʏɘ ɿɘʜ ni ɘǫɒmi ɿɘʜ ƨblohɘꓭ
Reflected. Thus did Venus rise	ɘƨiɿ ƨunɘV bib ƨuʜT .bɘƚɔɘlʇɘꓤ
From out the sea.	.ɒɘƨ ɘʜƚ ƚuo moɿꓞ

In order for parity not to be conserved in a process, it must be *impossible* in the sense that there is evidence that what you see in the mirror could *never* be made to occur in reality.

P   If mirror writing is not unnatural, then what is?

S   If you were to hold a ball in your right hand and let go, the ball would drop to the floor. In the mirror it would be a *left* hand which drops the ball, but the image ball would also drop toward the floor. Here the mirror corresponds to reality. However, if someone were to discover a special kind of ball which falls *down* whenever it is dropped from the *right* hand, but actually falls *up* when dropped from the *left,* this would break the symmetry.   Since the mirror image of this ball falling *downward* from the *right* hand would be a ball also falling *downward* from the *left,* the mirror would then contradict physical reality, namely, the fact that such a special ball always falls *up* when dropped from the left hand.

Fig. 2–1 *"The ball which falls up," or how the workaday world might look without parity conservation.*

P   I should have thought such fantasy more suitable to my profession than to yours.   Don't tell me a physicist found such a ball.

S   As a matter of fact, in 1956 two physicists named Lee and Yang conceived an idea which was in a sense equivalent to finding such a ball, and which subsequently obtained for them the Nobel prize.

The proposition may be stated as follows: Is there any experiment or any action whose mirror image is literally impossible to obtain in nature? When a person sees such an image, he should immediately realize that he is looking into a mirror, since no such event could ever actually occur.

Lest the reader underestimate the significance of this game with the mirrors, let us restate the problem in yet another form. It is now considered likely that this galaxy is populated by life forms other than our own, and that some of these may reasonably be expected to be rather advanced civilizations. It is not far beyond our present technology to develop transmitters which will beam electromagnetic signals to the stars. We can already communicate with the other planets in our solar system. By the same token we are constantly reviewing signals arriving from outer space for evidence of intelligent content. We may therefore with some justification postulate that eventually a successful communication will have been established. It is likely to be a rather distant star system, perhaps a hundred to a thousand light years away; and, since our signals travel with the speed of light, it will then take some hundred to a thousand years for a message to reach its destination. The messages will therefore probably be rather lengthy, in fact, continuous.

We may expect that after decoding and learning each other's languages we will commence an exchange of cultures. In addition to our scientific writings we may want to send them the best of our literature. The works of Homer, Shakespeare, and Dostoyevsky may be taking the long trip through space in the form of radio signals.

There is, however, one rather whimsical aspect to the whole procedure. The cultured residents of that distant planet may wish to read our works of literature in the original. But when they republish them after decoding our transmissions, the books will as likely as not appear in the form of mirror writing! This is due to the fact that when we describe our printed alphabet the message may say, for example, "The letter 'P' consists of a vertical stroke with a loop at the top extending to the right." But unless they know how to distinguish right from left, their version of a "P" is just as likely to be a "Я." In order to enable them to discriminate between a letter and its mirror image, we must somehow convey the difference between left and right. This is possible only if we tell them how to find something *which has no mirror image in nature*. Suppose, for example, that we discovered a universal natural law which makes it impossible for any living thing to have birthmarks or other asymmetries on the left side of its body, but permits them only on the right. Then we could use this as a means of telling everyone in the universe what we mean by the "right side." Before Lee and Yang no one in the world ever believed that such a thing could possibly exist.

Now this may appear to the reader as a matter of little consequence, and perhaps it is when considered of itself. But is it not rather strange that there is not a single detail about ourselves and our life on earth which we cannot successfully communicate, with the single exception of this difference between right and left? We can transmit radio photographs of all our great works of art, blueprints of all our machines, recordings of our favorite music; we can tell them how to duplicate every musical instrument, every bit of scientific apparatus. As a matter of fact we can, if their civilization is sufficiently advanced, direct them in building an exact scale model of our entire planet. But the cars in their model American city may turn out to have their steering wheels on the right side, and drive on the left side of the road, instead of the way we know it to be. All the books and billboards will then appear in mirror writing. Furthermore, we cannot even know for certain that this has happened. It is just as likely that they will have gotten everything "correct" as that they have it "backwards." This is the consequence of the parity principle, the fact that nature seems at first glance to have made no fundamental distinction between right and left.

The reader is invited to try to verbalize the difference between left and right. He will find it most difficult. Left is where your heart is? Who knows on which side an alien creature wears his pump? It will not take you long to convince yourself that right or left is a concept you can communicate only by gesture, or with a picture. If you can actually find a way to designate your right side by reference to any universal phenomenon, you probably deserve the

Nobel prize. This parity between right and left was accepted by everyone because it seemed as if one could always reconstruct any apparatus in a mirror image of itself and obtain the same experimental result. It makes no difference whether you stir your coffee clockwise or counterclockwise; it cools just as rapidly.

Suppose now that we are tired of cluttering up the radio communications channels, and have decided to send a space ship on the long journey to this distant planet, with a copy of every book which has been published on earth. This represents a considerable investment, and it will be most embarrassing if after hundreds of years in space all these books are delivered to people who have learned to read our language backwards. They may be too civilized to exhibit their anger and frustration openly, but the gist of the message they send back will be, "Why didn't you tell us that everyone on earth uses mirror writing?"

Lee and Yang actually found a way to avoid this predicament. In so doing they solved both formulations of the problem. They gave us an object, or a situation, whose mirror image has no counterpart in the real world; and as a result they also found a means of distinguishing left from right.

There are four known basic forces, or interactions, in nature. These are (1) the strong interactions, which hold protons and neutrons together in the nucleus of an atom, and which produce nuclear energy; (2) the electromagnetic interactions, some hundred times weaker, which hold atoms together, and are responsible for chemical and biological processes; (3) the so-called "weak interactions," which arise in radioactive decay and are far weaker than both of these; and finally (4) the gravitational, weakest of all the interactions between particles, but which hold planets and stars together only because these bodies are so large.

It was always assumed that parity is conserved in *all* interactions. We know that everything we have ever seen in a mirror appeared plausible, never absurd. It was one of the many "obvious" facts of life and only proves that physicists are not so very different from poets when forced to rely on intuition alone. Like the ball which never falls upward, all the evidence ever observed proved what everyone believed, until in their famous paper in the *Physical Review* of 1956 Lee and Yang proposed an experiment which was to prove that parity is *not* conserved in the "weak interactions" of radioactive decay.

This is the experiment; it was first performed successfully in 1957 by Mme. Wu and collaborators. It is fairly simple conceptually, although if the reader wishes to carry it out, he needs a sample of radioactive cobalt-60, and the means for cooling it to temperatures only a few degrees above absolute zero. The diagram shows one apparatus on the left, while on the right is a similar experimental setup, except that it has been built with everything inverted to make it *look like* a mirror image of what is on the left. (It is easy to see in the picture, however, that it is not *acting* like a mirror image.) If the reader's experience is such that he finds the thought of physical apparatus forbidding, he may wish to skim or omit the next paragraph. He should not in any case be able to predict the result of this experiment. Prior to 1954 no physicist on earth would have predicted it.

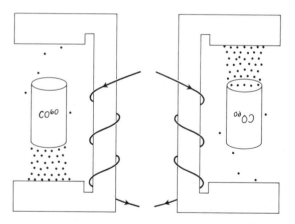

Fig. 2–2 *Parity is not conserved.*

The "C"-shaped structure is an iron electromagnet which becomes magnetic when electric current flows through the wire in the direction shown by the arrows. The tiny circular dots represent electrons which are normally emitted by cobalt-60 as a result of its radioactive decay. The surprising thing is that the effect of the magnetic field is to cause most of the electrons in the left-hand figure to be emitted in a preferred direction, namely, *downward*. In the case of the "mirror image" electromagnet on the right, however, most of the electrons travel *upward*. Now if they behaved as they should in a mirror image *these* electrons ought also to go downward. A "mirror image" apparatus fails to produce a mirror image of the result, and therefore a mirror image of the entire process cannot be reproduced in nature. If you perform this experiment in front of a mirror, the reflection in the mirror is something you can never see in real life!

Now we know how to communicate right-handedness. We tell our friends on the other planet to perform this experiment, then to slip one hand (or paw) inside a loop of the wire so that it is between the wire and the magnet, palm outward, fingers straight, with the thumb along the wire and pointing in the direction of the electric current. If the fingers point in the direction in which most of the electrons are emitted, it is the right hand. If the fingers point in the opposite direction, it is the left. It works for both sets of equipment, the one on the left and the "mirror" equipment on the right; thus it ´ es not matter which one they happen to build. We have therefore broken down the only remaining communications barrier between ourselves and the aliens.

Unfortunately there is just one thing which can go wrong. It happens that for every particle in nature there exists something called an "antiparticle," which has the same properties as the original particle, except for the fact that the electric charge (if any) is precisely opposite in sign. Thus the antiparticle of the negatively charged electron is the positron, and the antiparticle of the positively charged proton is the negatively charged antiproton. No one had ever seen an antiparticle until a theoretical physicist named Dirac solved the equation which bears his name, and obtained the prediction that such particles can exist in nature. It was a strange conclusion, and the apparent nonexistence of such particles was cited by Pauli, another famous physicist, in casting doubt on the validity of Dirac's work. While Pauli's criticism of Dirac was still in the process of being published, Anderson discovered the positron. We now produce antiparticles at will in our high speed accelerators. However, their life in this world is very short, since as soon as a particle encounters its antiparticle they are both annihilated, and their mass is completely converted into electromagnetic energy, or light. It is in fact this property of mutual annihilation which uniquely identifies the particle-antiparticle pair.

It has been speculated that there may exist whole worlds or even galaxies composed of antiparticle models of the atoms and molecules found in our world. There is no way for us to know whether what we see in the sky is a star or an antistar, since the light would be exactly the same in either case. Now if our correspondents should turn out to be antipersons living in an antiworld of antiparticles, we would be in real trouble. Our weak interaction experiment, in which they will use anticobalt-60, will give precisely the wrong result for right-handedness, and our book shipment will go on its way with thousands of books printed in what they would call mirror writing.

But this is the least of our troubles, because when our space ship finally makes contact with them, particles will meet antiparticles, there will be a big flash, and books and all will be annihilated, together with a piece of their antiworld. This is not likely to promote galactic friendship. However, until we find some actual evidence of stable antimatter in this universe, it would appear to be a bit paranoid to worry about this likelihood, so that we may assume that everything will be all right. In any case, we now know that nature can actually distinguish right from left, and Lee and Yang have in effect given us the "ball that falls upward."

Incidentally, the reader has without realizing it been introduced to yet another abstract kind of "symmetry." If we think of a mirror as a device for *transforming* a right hand into a left hand, then let us also imagine a magical device which changes particles into their antiparticles. Just as we wondered what happens when we invert everything to look like a mirror image, we now ask what would happen if while we were asleep a genie came along and transformed all the particles which make up an experiment into their antiparticles. Would we upon awakening be able to tell the difference? If, following such a transformation, everything behaved "normally," i.e., particles which had been moving upward continued to move upward, etc., then this would represent a symmetry. Suppose, on the other hand, the experiment involved a process which behaved one way for particles and another way for antiparticles. Then upon awakening we would rub our eyes and say, "Stop! Someone has been monkeying with my equipment! It is not behaving the way matter composed of 'real' particles should behave. That is not an experiment! It is an antiexperiment!" This would be an example of broken symmetry. In such a case the original experiment could never be precisely duplicated with antiparticles, just as in the case of parity breakdown a mirror reflection would represent a picture of something never found in nature.

This new symmetry is called *charge conjugation invariance,* to express the fact that in this type of transformation all electrical charges change sign; the negatively charged electron is replaced by a positron, etc. Thus symmetry in the physics sense is a way of asking a question. We say, "Suppose someone were to make a certain consistent change in a system, and then demand that, except for this single consistent change, everything must behave as formerly. Could the resulting situation actually exist in nature?" The parity mirror changes around right and left. Lee and Yang succeeded in breaking this type of symmetry. But in doing so they simultaneously broke charge conjugation invariance, because we happen to know that the replacement of all particles by their antiparticles in this experiment produces a different result. Thus we now know that neither of these symmetries exists independently in the weak interactions.

However, the idea of symmetry is not to be abandoned readily. It is far too attractive in helping us to build simple conceptual models of the physical world. Therefore it is not surprising that after the physicists recovered their equilibrium their first reaction to the downfall of parity conservation was to ask whether perhaps the combination of parity reversal *and* charge conjugation might not in fact constitute a universal symmetry of nature. This would appear to be the case in our cobalt-60 experiment. If the apparatus on the right, which was built to look like a mirror image of that on the left, had all its particles transformed into antiparticles, then the nature of the process is such that the anticobalt-60 sample would emit antielectrons *(positrons) downward,* instead of *electrons upward.* In other words, it would act just like a mirror image, but with particles replaced by antiparticles. Thus everyone breathed a sigh of relief and concluded that the symmetric picture of the universe was really not so wrong after all. It is just that the "universal mirror" which nature wants us to look into not only changes *right* into *left,* but also *matter* into *antimatter.* All events seen when we "look" into such a "mirror" are actually realizable in the physical world. A greater symmetry had been found, and Winnie the Pooh and Piglet had been vindicated.

Unfortunately this reassurance was destined to be short lived. Even the combined *CP symmetry* (charge conjugation and parity reversal) was soon broken in another weak interaction involving the decay of a particle known as the *K*-meson. This experiment (carried out by Christiensen and his colleagues in 1964) would not be reflected "properly," i.e., into something physically realizable, even by a mirror which also replaces particles by their antiparticles. Lee and Yang have apparently opened a Pandora's box in bringing down one of the pillars of the physics establishment, the conservation of parity.

Present theory predicts that the breaking of CP symmetry means that yet another symmetry, known as time reversal invariance, will likewise be broken. Experimental physicists are now busily searching for evidence of this. But this is the subject of our next chapter. The reader is free to speculate on the significance of all this. Perhaps if antiworlds exist they are populated by antiphysicists who write poetry and antipoets who wonder about symmetry.

Poet What led Lee and Yang to propose such an experiment? Was it just a hunch? Surely they weren't planning to ship books into space.

Scientist Actually they were trying to solve a problem in particle physics known as the *tau-theta* paradox. The *K*-meson was apparently decaying in two different possible ways, called *tau* and *theta,* which contradicted the known theory of this weak interaction process. It occurred to Lee and Yang that if parity were not conserved, this would resolve the paradox. They then looked through the records of previous experiments, and found that all the evidence for parity conservation involved *other* interactions than the weak. They therefore postulated that perhaps parity is not conserved in *all* weak interactions, and proposed experiments to test this hypothesis.

P   When you tell the aliens to do the experiment, how can you be sure that the electrons in their world will behave just as they do here?

S   There is a great deal more we know about this process than you have been told. This type of radioactivity is called beta decay, since electrons were once called beta rays. In addition to the electrons there are other emitted particles called neutrinos, which you failed to see in the diagram because they are invisible. They are also massless, they travel with the speed of light, and they spin as they move forward, much as a screw rotates when it moves forward under the action of a screw driver. Just as there are left-handed screws and right-handed screws, so there are left-handed neutrinos and right-handed antineutrinos. One type is produced when cobalt-60 decays, and the other when it is anticobalt-60. These are what give the process its "handedness." We also know from spectroscopic analysis of the light from stars that the same fundamental processes take place as for our star system. The only thing we do not know is whether it is a star or an antistar.

P   It was an interesting discovery. But I wonder if the game is worth the candle. Physicists have learned that parity is not conserved in weak interactions, and astronauts have walked on the moon. Meanwhile the rest of the world is unsuccessfully coping with the problems of poverty and war.

S   Actually, symmetries are not the exclusive province of artists and scientists. The very process of juxtaposing and examining real or supposed differences between individuals, groups, parties, races, nations, is itself a search for symmetry or lack of symmetry. I believe one can learn from the experience of the physicists in handling their "simple" problems. The symmetry property is a logical way of formulating a question. In the problems of our cities, for example, it is black-white symmetry; when we wonder about the role of women in society, it is male-female symmetry; in international politics it is east-west symmetry. Is it really true that Chinese are so different from Americans that they "don't mind" losing half their population in a nuclear war? Does oriental psychology differ from occidental psychology? How would a black society in America "enforce law and order" in cities filled with white rioters?

A physicist may not ask precisely the same questions a sociologist asks, or a politician, or an educator. But scientific method suggests a style of reasoning which demands precise formulation of a problem, and exposes preconceived ideas to objective and hostile scrutiny. This type of questioning is not conducive to the stability of a tyrannical government. A people which has been educated to scientific reasoning is, I believe, less inclined to accept rule by dogmas and slogans; nor is it easily persuaded to fight holy wars.

P   Then how do you explain Nazi Germany?

S   Technology flourished, but I would not call the Third Reich a *scientific* nation. It collected some of the dividends of past science. In order to fulfil Hitler's ambitions for the German people, it was necessary to destroy science while preserving technology. German science has to this day not fully recovered from the damage it sustained under Hitler.

P   You simply choose so much of history as supports your case and reject the rest. Physics may not itself be the fabrication of bombs, but it is undeniable that physicists have been and are involved in making them. And how do you account for the fact that Soviet scientists are among the best in the world in spite of having remained completely subservient to the state?

S   As a matter of fact, that is not quite true. The situation is more complicated. On the one hand, biology suffered a severe setback during the Stalin-Lysenko period; on the other hand, if you actually talk to them or read the Sakharov paper instead of the official literature, you will discover that Soviet scientists are not so likely to be found in "the silent majority" as you may have thought.

P   I fail to see any consistent improvement in the human condition resulting from your type of analysis. It may even be precisely the other way around. Every time you make a discovery or take a step to advance your field, you like to think you are improving man's state. And all you do is strengthen the hand of an establishment which turns everything to its own advantage. It makes no difference how beautiful the experiment, or how elegant the style of reasoning, it always ends up as bombs and napalm.

S Would a return to ignorance solve the problem?

P No, but we could stand a change of emphasis. The planets and the elementary particles have been around a long time, and they can wait a little longer.

S You mean the conquest of nature should be forced to come to a halt until men have solved all their social problems.

P You are so preoccupied with the conquest of nature that you fail to realize that people are a part of what is being conquered. You are so busy spraying the insects with DDT that you can't take time out to notice that the birds have stopped singing.

S What do you propose? Shall we join the flower children and look for "expanded consciousness" through drugs while we await our extinction?

P I wouldn't underestimate the hippies. They knew instinctively that the only way to avoid helping the establishment is to get outside it. As long as you try to function within it you end up becoming part of it. There is a place for everyone, even the dissenters. The idealistic white social worker enters the black ghetto to "help" impoverished people, and the establishment uses him to reduce unrest and keep the people from throwing rocks through its windows.

S There is something very false about this argument of futility. Withdrawal from society is just another luxury of the affluence which is supposedly being renounced. In order to bemoan the loss of microbes which have so zealously guarded the balance of nature you have to be born in a world without smallpox and plague. In order to develop a contempt for material wealth it has to be all around you. You turn on your stereo system and lie back dreaming of the days of the noble savage. These people who affect the antistatus symbols, the fetishes of the studied unkempt look and the torn jeans—do you believe their behavior actually conveys a message or represents a viable solution?

P   They are after all very young, and not always articulate. How else are they to tell you that they have rejected the values of your siren song of progress?

questions for the reader

1. If a mirror turns right into left, why does it not also turn up into down?

2. Consider an "Amazon" society in which the traditional social roles of men and women are interchanged. What symmetry-breaking processes would occur?

3. Why is it not possible to designate which is the right hand by transmitting a radio photograph of a hand?

4. If Russian troops had entered North Vietnam to suppress "counter-revolutionary" guerilla forces supported by the South, how would the Southeast Asia crisis have differed?

5. If tonight when you are asleep the entire universe is subjected to an inversion in which right becomes left and left becomes right, is there any way in which you will be able to notice the difference in the morning?

6. How do black nationalists differ from white supremacists?

7. Suppose we can determine exactly what the constellations look like as seen from the planet with which we are communicating, and can therefore designate right and left to the aliens by referring to these constellations. How may we then determine whether they are composed of particles or antiparticles?

8. If a European country which is a member of the NATO alliance were to choose a Soviet form of government and institute censorship of the press, would the "domino" theory require the United States to invade it and restore "freedom" at the point of a gun?

3 Time reversal

If you can look into the seeds of time,
And say which grain will grow and which will not,
Speak then to me, who neither beg nor fear
Your favours nor your hate.

WILLIAM SHAKESPEARE
Macbeth, Act I, Sc. 3

Physics is engaged neither in the development of time machines nor in the fabrication of bombs. But it is the business of physicists to take flights of fancy which carry them far beyond the boundaries imposed by current technology. When such "trips" take on a purely abstract character, when they are conducted in a systematic way, and when they begin to be organized into group tours, they are called mathematics. The nice thing about mathematics is that instead of contending with the hostility or indifference of nature, one is able to make up all the rules oneself. But once the rules have been laid down, one is honor bound to observe them. In mathematics such excursions are an end in themselves, but in physics they have the restriction that they must periodically renew their contact with the real world. It is a way of making rapid and inexpensive penetrations into regions of space and time which are physically still inaccessible, with a view to developing one's intuition and eliminating blind alleys. Eventually the experimentalists must follow up this type of reconnaissance with hard evidence to support what would otherwise remain only a conjecture. Physics is an experimental science, and all new ideas must be tested in the arena of observation.

Einstein liked to use what he called the *Gedankenexperiment*, the thought experiment, as a means of extracting information from nature. If science budgets continue to be reduced, perhaps we shall find ourselves increasingly dependent on this type of low-cost investigation. But progress will certainly eventually come to a virtual halt unless these psychic trips are followed up by physical experimentation.

We carried out such a *Gedankenexperiment* in the last chapter, and we shall have further recourse to this device in what follows. The symmetry known as time reversal invariance may be a bit more difficult for the reader to grasp, but if he has read this far he should be up to the task.

A process is symmetric under time reversal if it could also occur in nature with time running backwards. We do not actually have to turn time backwards to test for this symmetry; we can do it in our minds. If we have ridden on a train from New York to Washington, it is not difficult to see that the wheels could also run backwards and make the trip on the same track in reverse. Thus we say this process is symmetric under time reversal; a moving picture of the journey could be shown backwards without representing an impossible scene. The same is true of the earth's rotation about its axis, or its annual orbit around the sun. Dawn could have been designed to break in the west instead of the east, and winters and summers could likewise pass "the other way" without anyone's being the worse for it. These phenomena are thus symmetric under time reversal.

Physicists have long been intrigued with the question of the "proper" direction of time. If we are simply characters in a play written by the gods, as the ancient Greeks believed, someone might decide to turn the crank the other way, and then we would all encounter death first and birth afterwards. In fact, how do we know that this is not the "right" direction, and that someone has not already reversed it?

The currently accepted "big bang" version of the origin of the universe has us beginning with all matter concentrated in one region of space, followed by a huge primordial explosion in which stars and galaxies were created. This is consistent with evidence that the universe is presently expanding outward, and the stars growing farther and farther apart. However, such a picture raises a symmetry question with respect to the time boundaries of the process.

If there was a beginning, why should there not also be an end? Like an explosion in which the fragments rise into the sky and fall back to earth, perhaps the universe will eventually complete its expansion and commence falling inward toward a giant "implosion" back where it all started. If during the fall backwards time is somehow slated to run the other way, perhaps we are destined to live our lives all over again in reverse. In fact, it could be a cyclic process, in which we are called upon to reenact our little dramas repeatedly, forwards and backwards, again and again.

But let us take leave of phantasy, and return to science. At first examination it would appear that there are serious obstacles to the concept of time reversal. If someone projects a lifelike image on a screen with the projector running backwards, and tries to convince you that this is a real world, you will begin to laugh at some of the absurdities appearing before your eyes. Most people with movie projectors have enjoyed this game. One need only see children putting their toys *on* the shelf instead of pulling them off, and it becomes obvious that something is wrong. And when they come out cleaner in the end than they were at the beginning, it seems the time has come to toss time reversal invariance into the waste basket, together with all the other beautiful theories which failed to meet the test.

Let us make a generalization about this. There is something about nature which makes it prefer *disorder* to *order*. Physicists have found a precise measure of the degree of disorder in a system; they call it *entropy*. They have even invented a law, called the second law of thermodynamics, which states that the entropy of the universe is always increasing. It means that "on the average" the world will be a more disorderly place tomorrow than it was today. This may appear to be taking a rather gloomy view of things, but all parents of small children have experienced the effect of this law.

What do we mean by "order" and "disorder"? A rigorous definition takes mathematical form, but for our purpose the best way to convey such concepts is by example. If someone decides to toss a coin on the floor, we would be hard put to know *a priori* whether it will turn up heads or tails. This belongs to the future, and such knowledge is normally denied us. We can instead say only that the probability of heads is fifty percent. To understand this, let us consider tossing a hundred coins. Now we are in the advantageous position of being able to predict the future; namely, if someone is to enter the room after the coins have been tossed, we can predict with some confidence that he will see approximately fifty percent heads and fifty percent tails. This is a prediction we were unable to make when only one or two coins were tossed, since in that case they might easily have been heads only or tails only. If now the number of coins is increased to a thousand, and then to a million, the accuracy of our prediction continues to improve. Thus the larger the ensemble of events the better becomes our ability to predict the outcome. We say that we are approaching a "statistical sample."

Suppose that one million such coins have been tossed. The reader should be able to predict the outcome fairly accurately, but of course he could turn out to be wrong. If the experiment is repeated enough times, it is conceivable that the day will eventually come when all million coins will turn up heads! But we say the probability of this is extremely low, whereas the probability of fifty percent heads is relatively high. Let us call this maximum probability condition, with half the coins heads, the state of maximum "disorder," or maximum entropy.

If now a housekeeper enters such a room, in which the floor is strewn with randomly scattered heads and tails, and has been previously instructed to "just tidy up a bit," what should she do? One possibility is that she will selectively turn over all the tails, thus arranging all million coins with their heads up. We could then say that she has instituted a state of "order," or low entropy. This condition is one which if nature were left to her own devices would be extremely unlikely. Suppose that after the housekeeper has left, a hurricane lifts the house and all its coins into the air, causing them to tumble many times before settling to earth again. One will again find that approximately fifty percent of the coins are in the "heads up" state. This is what is meant by the statement that nature prefers the state of disorder to that of order.

But the physicist is not content with this observation. He also asks why it should be so. What is so special about the fairly equal distribution of heads and tails which causes nature to choose it and reject the state with all heads? The answer is simple, if not obvious. If one were to label all the coins with numbers from one to a million, and tabulate all the different combinations of heads and tails which can occur, it would be observed that there is only one combination which gives all heads, but a very large number of combinations associated with the fifty percent distribution. For example, the arrangement in which precisely half the coins are heads can occur if coin 1 is heads and 2 tails, or vice versa; or if both 1 and 2 are heads, and 3 and 4 are tails, and so on. It is simply that there are so many more ways for this "disorderly" state to occur than the "orderly" one in which all of them are heads which makes nature "prefer" it.

The more specialized version of the second law of thermodynamics is the statement that heat always flows from a hotter body to a colder one, rather than in the opposite direction. If we place two such bodies in contact, the combined system is highly "ordered," since it consists of a hot region in which the molecules are in an agitated state, and a cold region in which there is relatively less molecular motion. This is like having all our coins in the left half of the room heads, and in the right half tails. When the room is shaken up, it will finish with half heads and half tails in both sections. Thus nature "prefers" the uniform temperature conditions and the uniform coin distribution. Hence the more general statement of the second law of thermodynamics is that the disorder, or entropy, of the universe is always increasing.

As another example, consider the proverbial story of the monkeys at the typewriters. If one could hire enough monkeys and seat them all at typewriters, then sooner or later one of them would type all of Shakespeare's works with every detail of punctuation precisely correct. But if someone were to hold up an edition of Shakespeare which he claimed had been obtained by this method, you would be frankly skeptical.

Then what about the housewife who is busily tidying things up, washing the children's faces, returning their toys to the proper shelves? Is she breaking one of nature's "laws"? As a matter of fact, people are always struggling to achieve order; but even when they succeed, they pay a price in the form of heat and waste dissipated by their bodies. They may increase the order in one little corner of their world, but the net effect on the universe as a whole is to increase the entropy. In order to demonstrate this precisely, it would be necessary to introduce the reader to a more rigorous definition of disorder and entropy. But for us it is sufficient to note that it is this increase in entropy which the observer or movie viewer recognizes when he assures himself that time is indeed running the "right" way, and that no one is playing with the controls. Thus Eddington was led to call entropy "the arrow of time." Entropy points the way, "telling" time which way to run.

All this seems to be making a very bad case for the symmetry we call time reversal invariance, which predicts that processes can run *either* way in time. And yet I am forced to inform the reader that physicists have yet to find a single process which they know cannot actually run backwards in time if afforded the proper circumstances.

What is meant by "running backwards in time"? Everyone knows that clocks run "forward in time." Time reversal invariance implies that a watchmaker can design a perfectly good clock to run "backward," i.e., to act just as a "normal" clock would in a backward running motion picture. The hands of such a special clock rotate in the "counterclockwise" direction. We can then contemplate this clock alongside a conventional clock, and say that if the latter is running "forward," the former is running "backward."

But now let us try to design a clock in such a way that no similar clock could *ever* be made to run backward. If we can succeed in producing such a clock, then no watchmaker will ever be able to duplicate its operation in a time-reversed model. Such an "irreversible" clock might consist of two rooms with a small connecting door, as shown in Fig. 3–1. One of the rooms has been pumped full of air at an elevated pressure, while the other is at normal atmospheric pressure. When the door between them is opened at time "zero," the compressed air proceeds to leak into the other room for a while until both rooms reach the same pressure. A gauge in the high pressure room may be calibrated so as to measure the passage of time during this flow of air by reading the drop in pressure.

It may be difficult to market such a "clock," because it runs down so quickly and has to be "rewound" by pumping the air back

Fig. 3–1 *The clock which never runs backward. (What, never? ... Well, hardly ever.)*

into the high pressure room. But it is a clock of sorts, and if we now ask the "watchmaker" to build us one which will run "backward" in time he will ask, "What do you mean?" And we will say, "Make it look like the first clock would in a movie with the projector running backward."

Poet Time reversal is an imposing name, but I can't see any difference between this and mirror symmetry. In a mirror the rush of air going from left to right appears to be going from right to left, which you would call going "backward in time."

As a matter of fact, there is a connection. But if you look carefully, you will see that there is also a difference. We have installed pressure gauges, and even in a mirror the pressure will be seen to be *falling* in the high pressure room, despite the fact that right and left will be interchanged. Time reversal would require the pressure to be rising in the high pressure room, and falling in the low pressure room.

In any case, the "watchmaker" will scratch his head, because he knows that air will not pass from a low pressure region to a high pressure one. Does this mean that we have in fact broken time reversal symmetry? It would appear so, but now let us look at the process microscopically.

Consider the scene as it appears to us if we are traveling with *one air molecule* in the *low pressure room*. We are in constant motion, colliding with the other molecules and with the walls, and changing direction after every collision. As long as the door remains closed, we are just as likely to be going in one direction as another. But when the door is opened a new situation develops. Whenever we move in toward the open doorway, the frequency of collisions is increased because of the denser air, i.e., greater number of molecules, in the other room, and we are rather more likely than not to find ourselves turned around by such a collision and heading back the other way. Passengers riding the molecules in the *high pressure room,* on the other hand, experience exactly the reverse situation. When they try to pass through the door they encounter a *reduced* air pressure (fewer molecules), meaning that there now becomes a greater likelihood that they can travel a long way before making a collision which turns them around. Thus more and more such molecules find themselves passing through the door into the low pressure room, whereupon they discover that conditions are not at all conducive for making a return trip, at least not until the pressure in the two rooms has finally equalized.

However, all this is strictly a matter of probability. There are some "lucky" molecules in the low pressure room which upon heading for the doorway just happen to avoid making collisions and manage to reach the high pressure room. Likewise there are "unlucky" molecules in the high pressure room which just seem to stumble from collision to collision when they try to pass through the doorway, and never reach the less populated low pressure region in the other room. Furthermore, if we repeat the experiment daily, there will be "good" days and "bad" days, "lucky" days and "unlucky" days, just as for a gambling establishment. It is conceivable that one day the players will break the bank, but it certainly does not happen very often. That is why our "clock" is fairly reliable, and the "watchmaker" says it will never run "backward in time."

But it is nevertheless the case that, "had we but world enough, and time," we should eventually see the day when most of the air passes from the low pressure room into the high pressure room, and the lucky monkey finishes typing the last of Shakespeare's sonnets. Therefore *in principle* even such a special "clock" is symmetric under time reversal. We have failed to break the symmetry.

But the practical side of us, as expressed in the second law of thermodynamics says, "Forget it. You would have to wait many times the age of the universe for such a rare combination." That is why heat, which is simply molecular motion, always flows from a hot body to a cold one. But any *individual* molecule, if it could be photographed and shown with the film running backward, would be seen to be behaving quite normally. For each molecule the principle of time reversal invariance *does* hold, and travel both "forward" and "backward" in time is possible. It is only the *probable* behavior of a large statistical sample which is constrained by the second law. Time flows in that direction which is defined by an increase in entropy. And thus the large collections of complex molecules which comprise living things are always observed to die *after* they are born. We may rest assured that the gods have not playfully reversed the controls. Nevertheless, every individual link in the complex statistical maze which makes up a human being or a roomful of molecules can apparently occur in reverse! Time reversal invariance is a symmetry which we have yet to see broken in nature.

Why then do physicists spend their time looking for a violation of this principle? I shall try to give a brief answer to this question, but it will impose a bit of strain on the reader.

It should be recalled from the last chapter that when parity conservation was violated, for a while it was hoped that at least CP symmetry, the combination of charge conjugation with parity inversion, remained a symmetry of nature. Then in another experiment even this joint CP symmetry was violated. Now there is a super-symmetry principle called the *CPT theorem* which claims that under the *combination* of charge conjugation *and* parity inversion *and* time reversal all physical reality must transform again into physical reality. In order to understand what this means, let us imagine a special magic "mirror" in which not only is right changed into left, but also all particles are "reflected" into their antiparticles, and, to top it all, time is caused to run backward, as in a reversed movie projector. The CPT theorem tells us that if such a "mirror" existed, it would necessarily "reflect" the cobalt-60 experiment into the "image" of a perfectly valid physical process, whereas we saw that an ordinary mirror displayed something which could *not* occur in nature.

The CPT theorem is a consequence of a branch of physics known as quantum field theory, which has been developed to explain the fundamental high energy processes in which particles are created and annihilated. Now that a decay process has been discovered in which CP symmetry is violated, the only way CPT invariance may be preserved is for time reversal invariance to be likewise violated, in order to cancel the effect of the observed CP violation. Thus we find ourselves in the awkward position either of having to find an experimental violation of time reversal, which has so far never been definitely seen, or alternately of having to conclude that there is something wrong with quantum field theory. But the latter is a theoretical structure which has worked with fantastic precision in predicting the results of those experiments which have been successfully performed, in spite of the fact that it produces some mathematical difficulties. Experimental physicists are therefore very busy these days trying to save the necks of the theoretical physicists.

Poet   I suppose if I were a theoretical physicist I might be worrying about my "neck." But since I am not I can't help wondering about the relevance of such questions.

Scientist   You are very pragmatic. Has anyone ever asked that question about poetry?

P   There is no reason to ask it about poetry, or art, or music, because they are by their nature intrinsically human expressions.

S   What you are telling me in the twentieth century is

"Know then thyself, presume not God to scan;
The proper study of mankind is man."

P   I would not tell a physicist what to do. But it appears to most people that while the house is on fire the scientists amuse themselves by doing experiments with mirrors and movie projectors. It is a question of priorities.

S   Our house has always been on fire. Every generation finds its problems overwhelming. But that does not mean that philosophy, or art, or science, or poetry must wait until disease, poverty, and war have been abolished. Besides, your approach is too linear and simplistic. If in 1895 you had observed Roentgen passing electric current through an evacuated tube, you would probably have said, "Why don't you stop playing with induction coils and work on something practical, like setting broken bones?" And no one could then have told you that three months later X-rays would become available in surgery.

     Pope's expression is very appropriate. The business of physics is "presuming God to scan." At every stage it appears as if the most important questions in physics have been answered except for a single elusive detail. Right now that "detail" is the nature of the ultimate building blocks of the universe, the elementary particles. And symmetries like time reversal are the clues. The answers to such questions seem ultimately to exercise a profound influence not only on technology, and therefore the form of our life on this planet, but also on our philosophical concepts. And there seems to be no way of predicting the consequences.

P   It may be a profound scientific question, but it sounds a little like wondering how many angels can stand on the point of a needle. Why is it important to know that one molecule is capable of going the wrong way, or that once in a million universe lifetimes all the molecules in a room may go the wrong way?

     Knowing that something can happen, even if it is extremely unlikely, raises the possibility that we might be clever enough to cheat a capricious nature, forcing her to make up her losses in some other way. We can stack the odds in coin tossing by designing an unbalanced coin, or by tossing it "unfairly." The second law of thermodynamics may make the children's faces dirty, but it does not stop us from cleaning them. What one molecule does by chance

I may be able to impose on the others by choice. Just as the house-keeper turned over the half million coins with tails up, restoring "order" to the room, one can introduce a device in the low pressure room which selectively turns around those molecules heading away from the door and sends them sailing back toward the high pressure room. Such a device is called a pump. But if you lived in a society which had not yet developed pumps, such a concept would appear to you to be phantasy, as would the notion of a refrigerator which forces heat to flow from a colder body into a hotter one.

Time reversal invariance has some startling philosophical implications. It asserts that all processes are reversible in principle. The difficulty in actually achieving a reversal lies in obtaining what mathematicians call the proper "boundary conditions" for coupling the time-reversed process to the external world. If a man falls out of an airplane, for example, his initial downward velocity is relatively small, but the final velocity he attains upon reaching the ground is quite large; the points at which these velocities appear, namely, the beginning and the end of the fall, are "boundaries" of the process. In order to reverse the fall in time, this high *downward final* velocity must be replaced by a high *upward initial* velocity at the same point on the ground. Only then can the man "fall" back up into the plane. Nature is not particularly inclined to provide boundary conditions in the form of such high (and properly directed) initial velocities; it much prefers the low ones. Hence there is generally far more danger of falling out of an airplane than of being propelled from the ground upward into one. But it is not impossible. It is just that nature will not supply the catapults; we have to build and aim them ourselves.

Now since we can construct catapults, why can we not "time-reverse" people, make their bodies work backward, bring them back to life after they have died—just as we can "time-reverse" simple structures like atoms and molecules by supplying proper boundary conditions? The answer is that in order for a dead man to travel backward in time, he would have to be subjected to a very unusual set of boundary conditions. The embalmer must replace the blood he has previously drained. The wastes and carbon dioxide must be carefully collected and returned the way they left. Every cell in the body must be regenerated by precisely the inverse of the biological steps which led to its destruction. Thus the man grows younger every day. He unlearns all he learned. His muscles recover their tone. The presbyopia which affected his eyesight in middle age vanishes, and he places his reading glasses back into the hands of the optometrist who gave them to him. He unmeets the woman he married, forgets her, goes backward through the identity crises of adolescence, disobeys his parents, resumes crawling on all fours, and finally returns feet first to the womb whence he came, with the assistance of the obstetrician who delivered him. Of course we may not wish to carry him all the way back to the beginning of his life. We may prefer to turn the "pump" off at some intermediate stage, permitting him to go forward again. Will he do a better job of it this time if we warn him of the mistakes he made on his last trip? Or will he send us packing, insisting on the right to the same mistakes or a new set of his own choosing? Once more the "order" of youth and health succumbs to the "disorder" of old age and death, as his body resumes its search for the state of maximum entropy.

Although this appears "possible" in the abstract mathematical sense, since we have so far been unable to find a single individual violation of time reversal invariance in such processes, what makes it impossible in practice is the overwhelming statistical odds we should have to overcome. In the early stages of revival from the state known as clinical death, it is only a matter of providing one or two boundary conditions. Perhaps the heart has ceased to deliver energy, so we pump energy back into it, thus briefly reversing the process. But only a few moments after a complex interdependent system breaks down, nature and the second law of thermodynamics are quick to claim their due in the form of increased entropy. The process of degeneration and decay soon passes the point of no return, and we are back to disorder again, like the coins on the floor —as likely to be heads as tails.

On the other hand, simple elementary objects like the electrons and protons of which we are comprised readily encounter the conditions for time reversal, and therefore travel "forward" and "backward" in time with equal facility. When they are combined into more complex molecular chains, the possibility of obtaining the conditions for reversal begins to diminish. And finally in extremely complex systems like man it is defeated by sheer weight of numbers.

For us entropy is indeed the arrow of time. Irreversibility is the price we pay for complexity. Yet every *individual* chemical and biological step in a man's lifetime seems to be capable of occurring in reverse. Experimental physicists all over the world are at this moment hard at work searching for an exception to the time reversal principle. If and when they find it (and the prospects seem to vary almost from day to day), it will most probably again involve the weak interactions of radioactivity, and not the chemical processes on which life is based. But since this universal symmetry is now in question, it is quite impossible to predict either the outcome or the significance of the result.

Poet I must say I find your picture of a man having his excretions thrust back into his body to see if you can make him run backward rather offensive. The dehumanized scientific attitude exercises a symmetry in the contemplation of elementary particles on the one hand and human beings on the other. In the odorless tasteless super-world of the future will the people finally be replaced by perfect foam rubber dolls with plastic-synthetic organs which can run in either direction? Death will at last have been conquered; of course life will also, but then one can't have everything.

Scientist Why take it so seriously when it is not even possible? It was only a *Gedankenexperiment* to illustrate a point.

P I imagine Hiroshima must have started out as just such a *Gedankenexperiment*.

S Do you find the transplant of human organs equally offensive?

P At least that has a constructive objective; it saves human life. Your experiments are done strictly for amusement; you seem totally disinterested in the social consequences of your actions. I have a picture of you looking at the push-button which will send us all to the final Armageddon and musing with itchy fingers, "I wonder if it will really work?"

S I take it then that you would have me stop contemplating the symmetries of nature, and instead concentrate on problems like air pollution; or perhaps I had better join the ranks of people searching for a cure for cancer? What you do not realize is that physics at a given stage of development is not always ready to tackle those human problems which appear to be most in need of solution. What you are asking is that we restrict ourselves to the application of *old* physics for solving current problems, instead of developing *new* physics. You're tired of living off the interest; you want to start spending your capital. If your ancestors had never taken time out to wonder about fundamental questions, you (and I) would today believe that cancer patients were dying because they had offended an evil spirit. And you would not have the science of chemistry for coping with pollution.

P   I would also not have the pollution.

questions for the reader

1. Dust has settled all over the furniture and floors of a house, when suddenly a gust of wind through an open window blows all the dust into one room. Has the entropy of the house been increased or decreased? Is this consistent with the second law of thermodynamics?

2. In planning your research budget you find that you can afford to hire only one monkey and one typewriter. What is the probability that the first letter he types will be *a*? What is the probability that after punching the keys twice he will have typed the word *an*? That after three strokes he will have written *and*?

3. Richard Feynman says that every positron in the universe is an electron traveling backward in time. What symmetry process is involved in this statement?

4. If a man could be forced to live "backward," proceeding from death to birth, would he "remember" the past or the future?

5. The use of certain drugs is known often to induce "expanded" sensations and experiences of a hallucinatory nature, which are apparently random, unpredictable, and sometimes recur long after exposure. Does this suggest that the effect represents an increase or a decrease in the entropy of the brain cells?

4   Search for an absolute frame of reference

Oh wad some power the giftie gie us
To see oursels as others see us!
It wad frae monie a blunder free us,
      An' foolish notion.

ROBERT BURNS
To a Louse

Erich Fromm, in *Escape from Freedom,* advances the thesis that much of the tension of modern living arises from the sudden rush of freedom which has engulfed us—a freedom which most of the race actually fears and consciously rejects. From time immemorial it has been an important source of comfort for human beings to believe that every question has an answer, and that the way to obtain this answer is by reference to the proper authority. It is certainly far easier to locate an expert than to answer a difficult question oneself. And if all our friends consult the same expert, how secure indeed may we then feel with the answer he gives us!

Until some few decades ago children throughout the world were being educated in accordance with the doctrine of authoritarianism. The infallible source of knowledge was a book, or a scroll, or a president, or the leader of the party. Education consisted largely of committing to memory the pronouncements of various sources of knowledge. Most mature adults who are alive today, however dissimilar their political or cultural leanings, are products of this type of upbringing. In the final phase of this era science was promoted from a role akin to that of court jester to a position commensurate with its newly demonstrated power. After all, one ceases to be merely amusing when his incidental byproduct turns out to be the ultimate weapon. Accordingly a new oracle was created, although its powers were carefully circumscribed. We now have "scientific advertising," and "scientifically chosen ingredients," and of course "scientific weapons." There was even an educational children's record which joyfully beat out the words, "It's a scientific fact!"

Actually science is the spiritual antithesis of authoritarianism. "Fact" is a legal word, not a scientific one. There are no facts in science, only observations, postulates, deductions, predictions. The new generation which is today hopefully in the process of attaining maturity is the first to live through an adolescence in which all the idols are plainly seen to be standing on feet of clay. The executive of one great nation is exposed in the act of purveying misinformation as a matter of pragmatic realism, and the public learns to accept dissemblance as an instrument of national policy. The leader of another is admitted by his successors to have been slightly paranoid in imprisoning or executing some millions of innocent people in the interest of public safety, all with the support and assistance of the governmental bureaucracy. A religious authority issues a document on birth control so manifestly out of touch with the realities of this century that his own church is immediately observed to be unable to give it serious credence.

One may say that similar events have occurred in human history —but never before on television, with cameras and floodlights and news commentators who asked pointed questions, all in an era of mass education. The young are being exposed before they have had time to become properly inured or corrupted, and some of them are sporting large quantities of hair, taking drugs, and refusing to be packaged in the little boxes prepared by their elders. In the psychological semantics of the older generation, these young people are experiencing an "identity crisis." Alternately, if we permit ourselves to borrow from the language of physics, they appear to be operating without a "frame of reference."

The adult world, pondering how to curb its own propensity for self-destruction, is horrified, perhaps not without cause, at the sight of such premature youthful depravity, and begins to wonder whether the course of events ought not to be slowed down a bit, perhaps even reversed. Too much freedom, too much democracy, too much liberal education, too much questioning. One may choose to complete the picture with a historical allusion to the decline and fall of western civilization, accompanied by Chinese barbarians standing in the wings, ready to take over.

It is an interesting exercise in historical perspective to attempt to deduce causal relations among the various apparently unrelated aspects of man's cultural development. Thus we may readily infer a connection between the growth of classical physics from Newton's day to the end of the nineteenth century, and the appearance of the Industrial Revolution; and this in turn must have exerted a profound influence upon the literature, art, and philosophy of the period. By the same token the reader is aware of the effect nuclear physics has had upon the geopolitics of the latter half of this century, and hence upon its cultural patterns. What is perhaps less often noted is the *direct* influence of physics upon contemporary philosophy, even without the intermediate role played by technology. The effect of scientific determinism, for example, is quite evident in nineteenth and early twentieth century western culture.

I am not prepared to document the causal relations between the philosophical findings of modern physics, as obtained in the first half of this century, and the sociological phenomena which confront us in the second half. This is perhaps better left to some future historian; the reader is, however, invited to make his own judgment. What we do know is that the beginning of the twentieth century witnessed the collapse in physics of the previously accepted belief that there exists an absolute or "preferred" frame of reference for observing physical events, replacing this with the theory of relativity; that this was followed over the next two decades by experimental evidence for the concepts of quantum mechanics, causing the abandonment of the established deterministic picture of the world, and leading Einstein to make his famous remark, "I cannot believe that God plays dice with the universe"; and finally that if the human race learned anything from these surprising events, it is that no principle and no philosophy, regardless of how deeply rooted in historical precedent or intuitive understanding, is so absolute or so unassailable as not to be questioned or even abandoned in the light of evidence of its inadequacy.

The developments of quantum mechanics will be the subject of a later chapter. Here we address ourselves to the concept of frames of reference. We begin with a *Gedankenexperiment*.

An airplane is flying from Los Angeles to New York, and the stewardess is serving drinks. The passengers, who happen to be of a scientific bent, are engaged in an argument about how fast she moves when she carries the drinks from the rear to the front of the plane. In order to settle the question, they have strung a tape measure along the aisle, and have provided various passengers with clocks which have been properly synchronized. Thus the distance she walks may be measured, as well as the time it takes. Dividing the distance by the time tells them that she is moving at a constant rate of one meter per second, and the passengers can settle back and go to sleep, secure in the knowledge and mutual agreement that they have the correct answer to their question.

However, the plane happens to fly past a series of balloons carrying scientists who are likewise making measurements. They have been observing the same stewardess, who passes this way regularly, and here also an argument has ensued about how fast she travels when serving drinks. The scientists have strung tape measures between balloons and have likewise synchronized their watches, so that they too can make careful velocity measurements. They record the precise instant she passes each particular balloon. (We have designed the cabin of plexiglass, so that the stewardess is easily visible to outside observers.) But they obtain a rather different result from that of the airplane's passengers. Since the plane happens to be flying past them at an air speed of 300 meters per second, they observe the stewardess' speed to be 301 meters per second, of which one meter is due to her walk up the aisle, and the other 300 to the movement of the plane.

A further complication is introduced by control tower operators in the various airports between Los Angeles and New York. These operators are likewise placing bets on the same stewardess, whom they have been observing for some time through excellent field glasses. (Since this is a *Gedankenexperiment*, expense is no issue, and we can supply everyone with the best of equipment, better in fact than anything yet invented.) The distances between control towers are precisely known, clocks again synchronized, and these people find the stewardess to be traveling at 311 meters per second, of which 300 meters is caused by the motion of the plane through the air, one meter by the walk up the aisle, and another 10 meters per second by prevailing tail winds, which are blowing the plane, stewardess, passengers, and balloon scientists past the control towers. The only ones unaffected by this wind are the control tower operators, who fortunately inhabit relatively rigid structures.

Fig. 4–1 Frames of reference.

The question I now wish to pose to the reader is this: How fast is the stewardess *really* traveling? Is it one meter per second, or 301, or 311, or none of these? All of the highly qualified groups of observers, their tape measures, and synchronized clocks constitute what are called *frames of reference* for observing the stewardess. Each group obtains a consensus on the answer to the question within its own particular frame of reference, and each is satisfied with its own result. Of course, since they are educated individuals, and the discrepancy has no political or religious overtones, it may be presumed that if they were brought together to discuss the problem, they would be capable of divesting themselves of their various parochial points of view. However, this still leaves us to answer the question. Are the control tower operators better equipped to judge the stewardess' velocity than the airplane passengers?

Let us carry the process even further. A visitor from outer space has been assigned an observation post somewhere in the solar system. From the vantage point of his saucer he sees the earth traveling around the sun once each year, and this trip carries it through space at a speed of some 30,000 meters per second. Just now the earth and the airplane happen to be going in the same direction. Thus the stewardess in passing this observer not only has her own motion, and that of the plane, and that of the wind, but also that of the earth itself. The creature therefore claims the stewardess is serving drinks at a speed of 30,311 meters per second. (We will neglect the further daily rotation of the earth about its axis.) But his distant monitor at Galactic Headquarters, knowing that the entire solar system, including sun and earth and all the planets, is itself likewise traveling through space, issues a report saying, "Outpost 4871 Sigma should be relieved at once. Parochial Identification Syndrome has set in, advanced stage. Loss of reference frame. Appears to have fixation on native airline stewardess, reports speed without accounting for local star movement." And the relief outpost will come carefully briefed to remember not only that the sun and stars are in relative motion but also that all the galaxies in the universe have been receding from each other since the day of the "big bang," so that, strictly speaking, velocity measurements must properly include the effect of such solar motion.

But how about the point from which the galaxies appear to be flying outward? Is this then finally the basis for an Absolute Frame of Reference? Or is it likewise moving with respect to yet another point, and so *ad infinitum?* Whom shall we believe? Is there such a thing as *absolute* motion through empty space?

For the answer to this question we consult no lesser authority than Sir Isaac Newton (1642–1727) himself. This remarkable individual nearly three hundred years ago established laws of mechanics and mathematics which in the nineteenth century made it possible to predict the existence and location of the planet Neptune before it had ever been seen, and in the twentieth century launched man into the space age.

Newton's laws of motion predicted that there could never be a mechanical means of measuring absolute motion through space at constant velocity. In other words, the passengers in our airplane would be hard put to know the speed of the plane without looking outside to see how fast it is moving *with respect to something else.* But the human race was not ready until the beginning of the twentieth century to abandon the security derived from belief in an absolute frame of reference. Just as there apparently had to be an absolute space reference, so there must likewise have been an absolute scale for measuring time. Newton wrote in his *Principia:*

> Absolute, True, and Mathematical Time, of itself, and from its own nature flows equably *without regard to any thing external,* and by another name is called Duration ... Absolute Space, in its own nature, *without regard to any thing external,* remains always similar and immoveable.

(The italics are mine.)

Newton, having noted that *actual* measurements of space and time must always be *relative* to something, nevertheless felt it necessary to decree into existence an *absolute* reference frame which would never be observed by anything or anyone! He might have known better than to engage in such tautology. For if something interacts with nothing, how can we meaningfully explore the question of its existence, even in principle?

It is interesting in this connection that such a logical trap was apparently avoided hundreds of years earlier by the philosopher

St. Augustine (354–430 A.D.), who was never to have the benefit of Newton's physics or mathematics. The disciples of St. Augustine are reported to have asked him the following question: What did God do before he created the universe? And the answer was: Before God created the universe, there was neither space nor time, so the question has no meaning.

But in 1865, long after Newton's death, a strange thing happened. It suddenly appeared to be possible to measure absolute velocity in empty space. An experiment was proposed which when carried out ought actually to enable us to determine how fast our airline stewardess is *really* traveling. For the first time it appeared that we could find out who has the Ultimate Frame of Reference, who is telling the "truth"—the passengers, or the balloon scientists, or the control tower operators, or the solar system outpost, or the galactic monitor, or the Point in Space where it all began.

5

Michelson, Morley, and the ether wind: the experiment that failed

He rode upon a cherub, and did fly:
yea, he did fly upon the wings of the wind.

Psalms, XVIII, 10

This is the way things stood: The stewardess was walking through the cabin; the cabin was plowing through the air; the air was blowing past the control towers; the control towers were moving with the earth; the earth was traveling through space. Therefore the stewardess was traveling through space. But at what velocity? Clearly an observer who was himself in motion could not be relied upon to give us the correct answer. It would have to be someone who was *not* in motion, and people living in the second half of the nineteenth century thought that at last they knew how to find such an observer.

The development which produced this state of affairs was a set of mathematical expressions known as Maxwell's equations. The phenomena of electricity and magnetism had been carefully observed and measured. Maxwell was able to condense these observations into a very beautiful theory, which was so much in accord with the way nature actually behaves that it has survived to this day, and is yet to be contradicted by any observed phenomenon.

But, as sometimes happens in such cases, his theory also proceeded to predict something completely unexpected. Maxwell's equations predicted a rather precise numerical value for the speed of light *in empty space!* Now this announcement may not at first glance produce or dispel any particular anxieties on the part of the reader, or make him want to turn cartwheels. But it ought to suggest that the airline stewardess *does* have a unique velocity after all (as one always suspected), that there *does* exist an Observer who can measure it, and finally that we are in a position to determine who this Observer is. All we have to do is ask our conflicting observers to measure the speed of light; the one who gets the right answer, i.e., that predicted by Maxwell's equations, will be telling the Truth. He is the Ultimate Authority, the Absolute Frame of Reference. All the other so-called authorities are telling a lie, or at least are in error because they are unaware that their reference frame is itself moving with respect to that of the Ultimate Authority, and that this motion will cause them to read all velocities incorrectly. When these false prophets try to pass off the wrong velocity for the stewardess, they may succeed in throwing up a smokescreen, because her velocity is controversial. But when they give us the wrong answer for the speed of light, then we will have them. Not only will we know that they are in motion, but we will even know their velocity.

Newton would have been most excited to learn that Absolute Space, which had appeared to him to exist "without regard to anything external," could now finally be identified. But the physicists of that day proceeded cautiously. It is one thing to ask all our observers to do a *Gedankenexperiment* in principle; it is another to provide them with apparatus for making real measurements. The physicists began first to elaborate on their newly discovered theory. Since light was known to be a *wave motion* (and this was consistent with Maxwell's findings), something had to be waving. In the case of ocean waves, it is the water which is waving; in the case of sound waves it is the air. But the light reaching us from distant stars is "waving" in empty space, which we call a vacuum. There was at the very least a problem in semantics. It was resolved by naming whatever it was that was waving the "ether," and this "thing" had the property that it existed "without regard to anything external." You could not see it or feel it or smell it; its only function was to propagate electromagnetic waves (such as light) with the speed predicted by Maxwell, namely, 300,000 kilometers per second (10,000 times as fast as the earth in its trip around the sun). Any reference frame, i.e., any set of observers and clocks and tape measures, which measured the speed of light and found it to have this "correct" value, must be "at rest" in the ether. All other frames of reference which are moving with respect to this one are "in motion," and must experience an "ether wind," causing them to measure a different value for the speed of light from the "correct" one predicted by Maxwell. It is the same effect experienced by a water skier who is being pulled through ocean waves by a motor boat. The waves hit him at greater speed than they would if he were simply floating, i.e., at rest in the water.

Let us return to our flight from Los Angeles to New York. We can dispense with the stewardess for the present. Let us also evacuate all the air from the cabin of the plane (after having first taken such routine precautions as providing the observer-passengers with space suits). We instruct a passenger at the rear of the plane to strike a match or turn on a flashlight at a prescribed time. Another passenger in the forward end of the cabin will record the precise moment when he first sees this light signal. The length of the cabin divided by the time it takes for the light to travel this length gives us the speed of light *in vacuo,* as observed in the airplane's frame of reference. If this turns out to agree with Maxwell's prediction, the airplane is the Absolute Frame of Reference and its passengers the Ultimate Authority for measuring absolute velocity through space. In the twentieth century, when authorities and oracles are crumbling all about us, we should surely cherish these people, since the answers they give us can be relied upon. If, on the other hand, they get the wrong answer for the speed of light, we shall consign them to the special limbo we have reserved for false prophets.

Poet It seems that you are testing the reliability of one authority by requiring him to agree with another. What makes you believe Maxwell's prediction for the "correct" speed of light? The physicists apparently have hierarchies of truths. Perhaps one day we shall discover that what we have called science is really only the worship of equations and theories which have been elevated to a preferred position by the scientific establishment.

Scientist Since all the scientists we know are human, they undoubtedly suffer from the weaknesses which beset the rest of the race. They have their establishments, their preconceived notions, and their authorities. But they never tire of testing accepted principles and "obvious" conclusions with new and ingenious experiments. When a discrepancy appears, and is substantiated by repeated and careful measurements, the entire community closes in for the kill. And no "authority" can on the basis of previous success alone expect to survive the critical climate engendered by an experiment which successfully upsets the established theory.

Maxwell did have the confidence of the physicists of his time. But it was not based on faith. He had earned it by a rather remarkable achievement. He had analyzed the behavior of electrically charged bodies and magnets, which appeared to have no connection with light itself, and had come up with a mathematical prediction which related the speed of light to the behavior of such bodies. Furthermore, experimental techniques in the nineteenth century were sufficiently refined to measure the speed of light, and the observed value agreed with Maxwell's theoretical prediction very closely, to within experimental error. But the speed of light is so great that the change introduced by the motion of the observer through the ether is too small to have been detected in these early measurements. What was therefore needed was an experiment which would effectively measure not the speed of light itself, but rather the *change* in speed produced when the observer travels with or against the ether wind. It is somewhat similar to the fact that it takes approximately five and a half hours to fly from New York to Los Angeles, and only four and a half hours for the return trip. It is the effect of prevailing winds, which blow generally west to east. There has to be a similar "ether wind," if the earth is plowing through the ether, and the speed of light would therefore have to be different "downwind" from what it is "upwind" or "crosswind."

It was sometime in 1881 that a physicist named Michelson first carried out his famous experiment to find the ether wind; he and Morley repeated it in 1887, and then many times thereafter. We shall not concern ourselves with the details. There were mirrors which sent light along different paths, and a heavy stone platform floated in mercury. What is important is that *they could find no ether wind,* not in the direction the earth was moving, and not in any other direction. Furthermore, they climbed mountains, they repeated it at different times of the year, they rotated the apparatus in various directions. Always the speed of light was precisely the same. Now the earth in its orbit around the sun has a speed of 30 km. per second, and the experiment was capable of detecting as much as 10 km. per second of motion through the ether. Thus failure to observe different speeds of light at different times of the year suggested that the earth must be "at rest" in the ether. It was therefore the "preferred" frame for measuring absolute motion in space. Yet we have known since Galileo that the earth is not the center of the universe. Why should it be at rest in space?

The experiment to observe the ether wind had failed. But the failure was not an inadequacy in experimental technique. It was a failure of nature to produce results which men could find reasonable. Let us translate this into the context of our *Gedankenexperiment* with the airplane. When the passengers measure the speed of light in their cabin, using their clocks and tape measures, they find that it agrees precisely with that predicted by Maxwell! They therefore announce to the world that their airplane has passed the test. It is at rest in the ether, and they represent the Ultimate Authority; theirs is the Absolute Frame of Reference. But while they are breaking open the champagne to celebrate their new exalted status, the balloon scientists are carrying out a similar test. They have, it is true, some trouble in evacuating the air in the region between balloons. (As it turns out, the air has no significant effect on the light signal.) But they manage to send light signals from balloon to balloon, and they measure distances and time intervals. To everyone's amazement, they too get Maxwell's predicted answer! Now we have two Ultimate Authorities.

The control tower operators, not to be forgotten, are certain that they are more "at rest" in space than a moving airplane or a floating balloon. And they proceed also to obtain the correct answer. Furthermore, the newly appointed solar observer, who has found himself a position somewhere between the orbits of earth and Mars, proceeds to confirm this result, finding light to travel with the same speed observed by terrestrial experimenters. And so throughout the universe. All observers agree with Maxwell's predicted speed of light, regardless of how fast they are moving or not moving.

In the event that the reader is not properly baffled by all this, consider the following: Suppose someone were to claim that all measurements of the stewardess' velocity, as obtained by the different observers, were in complete agreement. In other words, the passengers find her to be walking down the aisle at one meter per second; the balloon scientists say she is moving at this velocity from their viewpoint also, despite the fact that the motion of the plane should be adding its velocity to hers; the solar and galactic observers likewise claim to obtain this same value of one meter per second regardless of the movement of the earth and sun. The reader would say that this is clearly implausible, that it violates common sense. And he would be right. Actually, the velocity of the stewardess is, fortunately for the sanity of all of us, *different* as measured by the various observers. This paragraph is just a bad dream.

Yet what Michelson and Morley found was apparently just as preposterous. It simply did not involve the stewardess; it applied instead to the speed of a light signal. All observers, regardless of their motion with respect to each other, were finding light waves to move past them at the same speed. It is as if the water skier were to report that ocean waves are passing him at the same rate as that observed by a stationary swimmer floating on his back in the same water, despite the fact that the skier also has the additional speed of the motor boat which is pulling him. It simply did not add up. Why should the stewardess be traveling at *different* speeds for the various observers, while light travels at the *same* speed?

Poet I can see a difference in the two situations. The stewardess is *riding* the plane, and therefore *acquires* its additional velocity. But the light signals are started independently by each set of observers in separate experiments.

True. And to make the two situations correspond, we had better get them all to observe the same light signal. Since the cabin of the plane was built of plexiglass in order that everyone could have a good look at the stewardess, we shall take advantage of this. Let a passenger at the rear of the plane turn on the light signal as he passes one balloon observer; then another balloon observer farther up the line records the instant *he* sees this light, and divides the distance between balloons by the time interval. In this way we can get a fair test of how fast the *same* light signal travels in the balloon scientists' reference frame, as compared with that of the airplane passengers. We then tell the control tower operators to do likewise, and also our cooperative alien assistants out in space.

Actually, the Michelson–Morley experiment *does* use the same light source, and the result is still no different. All the observers find the speed of light to be the same. There is no escape. We are faced with the fact that nature is behaving "illogically"; it is adding one and one, and getting one. We shall shortly see that Einstein's response in effect was to shrug his shoulders and observe that this is the way things are. If nature refuses to do what we think it ought to, if it fails to follow the rules of logic we have laid down, it is *we* who must make the adjustment. We have uncovered a new world, and there is nothing for us but to learn to live in it. Of course we shall have to rewrite some of the rules, divest ourselves of some of our prejudices. But if we can bring ourselves to make the adjustment, this new world may prove to be quite livable. Who knows? It might even turn out to be better than the old one.

While the physicists were pondering what to make of all this, most of the world was going about its business completely oblivious to the problem of the ether wind. In the United States the big issue was Silver; William Jennings Bryan delivered his famous speech, "You shall not press down upon the brow of labor this crown of thorns, you shall not crucify mankind upon a cross of gold," and went down to defeat with the Populists in the election of 1896. Nicholas II of Russia was suppressing attempts at reform; cossacks and peasants were massacring Jews in the ghettos; and in 1903 a small group of Social Democratic revolutionaries met in secret and voted to support Lenin and the Bolsheviks in structuring a monolithic party apparatus. France was torn in two by a struggle between the Dreyfusards, who believed in a democratic republic, and anti-Dreyfusards, who believed in the authority of the army and the church; and Emile Zola in 1898 wrote his famous *J'Accuse*. It was the turn of the century, and physics was hardly the issue of the day. But the Michelson–Morley experiment, the experiment that failed, was a slowburning fuse which was to reach far into the next century.

6   The postulates of special relativity: Einstein learns to add

Our birth is but a sleep and a forgetting:
The soul that rises with us, our life's star,
   Hath had elsewhere its setting,
    And cometh from afar:
   Not in entire forgetfulness,
   And not in utter nakedness,
But trailing clouds of glory do we come
   From God, who is our home:
Heaven lies about us in our infancy!

WILLIAM WORDSWORTH
Intimations of Immortality

I once heard a high school mathematics teacher explain to a roomful of parents that he had two objectives: The first was to teach the students plane geometry, and the second was to teach them not to believe everything he told them. To my surprise the response of the parents was nervous laughter and derision. They were still living in the nineteenth century, despite the fact that they had all been born in the twentieth. Theirs had been a happier time. One knew where one stood. These are confusing times.

The problem faced by the young Einstein in 1905 was to explain the failure to find an ether wind. The old concepts had provided no explanation. According to the old rules, for example, a racing car must pass another car in the same race more gradually than it passes the spectators in the stand. Thus if the slower car 1 drives past the spectators with velocity v_1, and the faster car 2 passes them at velocity v_2, then the velocity u at which the fast car overtakes and passes the slower car is given by the difference

$$u = v_2 - v_1 . \tag{6-1}$$

Let us rewrite this as

$$u + v_1 = v_2. \tag{6-2}$$

But this simple rule of addition of velocities had broken down, as may be seen in Fig. 6-1, which shows car 2 overtaking 1. If we now remove car 2, and replace it instead with a light signal produced when someone suddenly switches on a flashlight, this light signal then races down the track after car 1. The Michelson–Morley experiment established that the universal speed (let us call it c) at which the light signal overtakes and passes the car is the same speed at which the light signal passes the spectators!

In other words, the above equation becomes

$$c + v_1 = c, \tag{6-3}$$

despite the fact that v_1 is not zero. Worse yet, it turns out that if driver 1 decides to soup up his engine to such an extent that his car is speeding along at practically the speed of light itself, this relationship still holds, becoming

$$c + c = c. \tag{6-4}$$

How does one explain such patent violation of common sense?

Einstein took a very bold step. He decided not to believe what his teachers had told him. If one and one is observed to add up to one, instead of two, there is nothing for it but to change the addition rule. Mathematics was invented by people; it can be changed by people. When we drop a ball into an empty basket, and then drop another ball into the same basket, and finally count the number of balls in the basket, we find that there are indeed two balls. Is this obvious? Not to a newborn infant. He believes it only after he has tried it several times. Then he proceeds to grow old and cynical, and claims that it was obvious.

The Michelson–Morley experiment was nature's way of telling us that the rule for addition of velocities is not the same as for balls dropped in a basket. The only reason this appears to us to violate common sense is that our idea of "common sense" is simply the summation of all the experiences we have had since infancy. It happens that before the time of Michelson no one on earth had ever made accurate measurements involving speeds as great as that of

Fig. 6-1 *Results of the Michelson-Morley experiment. The faster car 2 overtakes and passes slower car 1* gradually, *i.e.,* more slowly *than it passes the spectators. But a* light *signal passes this car at the* same speed *as it passes the spectators.*

light. Einstein concluded that the correct rule for the addition of velocities is not the old familiar one at all, but is merely *approximated* by the old rule when the velocities involved are the much slower ones encountered in everyday observations; however, when speeds become so large as to approach that of light, the new rule must be such as to predict what had now been observed, namely, that light has the same speed for everyone. He proceeded to look for a rule of addition of velocities which would meet these requirements. Clearly it could not be the same rule as for adding balls thrown into a basket. The derivation of this new rule, which is left to the Appendix, requires only the algebra we learned in high school, and is not very difficult. What was truly difficult was the decision to question the old rule!

Let us pause to take note of a very special property of the frames of reference we have been investigating. It is that they are all moving at *constant velocity* with respect to each other. This means that both their speed and their direction do not change. The Special Theory of Relativity, which we are about to contemplate, is restricted to such systems, unlike the General Theory, which is not. In other words, the pilot must fly in a straight line at a uniform speed, without bumps or jerks or tricky maneuvers. Otherwise the stewardess will drop her tray, and meter sticks and clocks go tumbling all over the cabin. The same is true of the other reference frames. The racetrack is perfectly straight during the period of the experiment, and the driver makes certain that his speedometer reading remains constant. Such reference frames are called *inertial*, because everything is observed to obey the familiar laws of inertia. On the other hand, if the pilot breaks the rules, and the plane goes into a power dive, an unexpected force will cause the passengers to pile up at one end of the cabin; and if the racing car leaves the straightway for a sharp turn, centrifugal force may yank the driver out of his seat, or the car off the road. None of these unpleasantnesses will appear in our *inertial* frame of reference. Each such frame has the property that its observers do not spill their soup, and when they look out the window at another inertial frame, they will see it passing them at constant velocity.

Such frames of reference are also known as *Lorentz frames*, after the man who first wrote down a mathematical expression known as the *Lorentz transformation*. Just as we saw in Chapter 2 that a mirror *transforms* an experiment into its mirror image, by showing us how the world would appear to an observer who sees right as left and left as right; and a movie projector running backward *transforms* it into its time-reversed image by showing us how it would appear to an observer for whom time is running backward; so a Lorentz transformation mathematically transforms an experiment into what it would appear to be to an observer in another Lorentz frame.

Einstein summarized the findings of the Michelson–Morley experiment in two very simple postulates:

Postulate 1 *All inertial reference frames are equivalent.*

This means that there is no "preferred" or absolute frame of reference, and hence no way to measure absolute velocity. In fact, there is no means for an observer to distinguish his frame of reference from other frames in any special way.

Postulate 2 *The speed of light is the same for observers in all inertial frames.*

The first postulate reaffirms the fact that the one meter per second at which the airplane passengers see the stewardess moving is neither more nor less her "real" velocity than the 311 meters per second claimed by the control tower operators, or the 30,311 meters per second observed by the galactic outpost. The control tower is no more "at rest" than the airplane. When the window curtains are drawn, the inhabitants have no way of telling whether they are in the tower or in the plane; when they open the curtains they can learn the velocity of the other frame with respect to theirs, but they can never determine the absolute velocity of either.

The first postulate takes us back to the way things were in Newton's day; there is no way to measure absolute velocity. And the second postulate says that even Maxwell's prediction of the speed of light in empty space affords no way to determine one's absolute velocity, since all observers riding inertial frames obtain the same speed of light, regardless of their velocity.

What is interesting about the second postulate is that it accepts as basic premise what everyone had been trying to explain away. People had been suggesting various mechanical explanations for the Michelson–Morley results; perhaps it was something in the apparatus, perhaps moving objects change their dimensions for some inexplicable reason, perhaps the ether is affected by the presence of the earth. None of these explanations had proved to be satisfactory. The one thing no one had been willing to consider was that the rule of addition taught him by his teachers might actually be wrong in this case. In the second postulate Einstein parted company with Newtonian physics. He said, let us begin with the experimental result and from this *deduce* the rule for addition of velocities; let us learn to add again.

What Einstein did was to interpret the Michelson–Morley results in the simplest possible way. But the postulates are no less profound for being so simple; quite the contrary. We shall shortly see how strange is the world into which they have plunged us. Also, there is an important philosophical distinction between Newton's view of things and Einstein's. Newton had said that there is no way to measure constant absolute velocity; Einstein agreed. But Newton persisted in asking himself whether such velocity can nevertheless exist, and concluded that it must, despite the evidence that it could never be observed. To Newton a universe without an absolute frame of reference was inconceivable. To Einstein, after the Michelson–Morley experiment, it was like the question asked St. Augustine by his disciples, and his answer was effectively the same—not that there is no ether, but that the question has no meaning. If we accept *in principle* that there is no way to observe the ether and its absolute reference frame, it profits not to ask whether it exists. The ether was a human invention; now it appeared that, like Absolute Space and Absolute Time, it was Absolutely Unobservable. A question which is so defined as to have no answer *in principle* is hardly worth the asking. And with this pointed omission Einstein freed the world of the ether web it had been spinning for more than a century.

Poet It is interesting that you like to identify these scientific developments with freedom. Science always comes bearing the gift of freedom, and we end up more enslaved than ever. Twentieth century man is a dehumanized cog in a culture dominated by machines. Stripped of any sense of values, morally and spiritually bankrupt, he waits at the edge of a precipice into which a politician or general will decide when to provide the final shove. And what you see in all this is freedom.

Scientist The precipice is another thing. But in the business about cogs and machines you are a bit behind the times. The Industrial Revolution is a product of the classical era in physics, which ended with the nineteenth century, and many of its problems have already been superseded. Naturally, we are still feeling the effects. But the new modern physics is part of a completely different world which we still do not fully comprehend. The youth revolt, the breakdown of authority, the anti-conformity, the wailing music, and the pop art do not look to me like products of dehumanized cogs; if anything, they are backlash.

P   Have you ever wondered what the world would be like if Einstein had kept those postulates to himself?

S   It is a useless question. If Maxwell had not deduced his equations, someone else would have. If Michelson had not built an interferometer, someone else would have done the experiment. If Einstein had not drawn the right conclusion, someone else would have had to see it. Each step opens the door to the next one. You might as well wish that Gutenberg had never invented the printing press, or that Socrates had never asked such pointed questions, or that some unknown Prometheus had never discovered how to control fire. If you want to protect yourself against human knowledge, where do you begin? The wheel? the stars? Shakespeare? Shall we burn the books?

P   It's a bit late for that.

7   Is meaning important?
The relative character of simultaneity

Of all small things
That have the most infernal power to grow,
Few may be larger than a few small words
That may not say themselves and be forgotten.

EDWARD ARLINGTON ROBINSON
Genevieve and Alexandra

When we speak of the polarization of the intellectual world into "two cultures," we are actually referring to a set of labels artificially imposed by a system which forces each individual to choose between "science" and "humanity." Although this may facilitate the preparation of professionals for specialized jobs, in the real educational sense it is a false dichotomy. No isolation is possible for the "cultures" themselves, for the simple reason that neither can exist without the other. It is certainly true that technicians (of both types) can be trained to solder wires or sell advertisements without much knowledge or understanding of the "other culture" (or of their own, for that matter). But this kind of pragmatic specialization is just a way of giving people time cards to punch, and has little to do with culture.

Some years ago an educational problem arose in the secondary schools in this connection. Those students who were to take physics by the time they were seniors had to be prepared by a suitable hierarchy of courses, and the question arose of selecting those with an aptitude for physics before they had taken any course in science. It was found that the best correlation with eventual good performance in physics was obtained by those students who had the highest grades in English.

This is probably not coincidental. Language plays an important role in physics. It is popularly believed that physicists communicate only in equations, and the comic book picture of the scientist has him standing over bubbling fluids and flashing lights as he searches for secret formulas to give him control over human destiny. Physics actually is built on words as well as equations, and I suspect that without the rich profundity of language which is our heritage from the writers, poets, and philosophers of the past, there would be no physics today. It is seldom appreciated that the imposing power of nuclear energy, and the host of problems it has created, can be traced (among other things) to a question about the meaning of a word.

The word is "simultaneous," and its definition is the subject of this chapter. It is strange that our ability to move mountains, send space ships to the stars, and most important of all, understand the nature of the universe, should hang on such questions. Yet this is the way things are.

What do we mean by "simultaneous"? "Obviously," says the reader, "we mean 'at the same time.'" And it is at this point that the physicist and the semanticist may have to part company. The latter is interested in how words have actually been used, whereas the former is concerned about possible ambiguity or meaningless-ness in the concept behind the word. We have already seen two examples of such confused usage. In one case Newton (a physicist) employed terms like "Absolute Space" and "Absolute Time," when he had no real way of defining them. In the other St. Augustine (a humanist) recognized that his disciples could not meaning-fully ask what God "did" before there was a space and a time in which to "do" it.

It helps to avoid confusion if one can define every questionable expression in terms of an experiment—actually a *Gedankenex-periment* will do nicely. Before the twentieth century the definition of "simultaneous" as meaning "at the same time" would have appeared unambiguous, because people were still clinging to the notion that there must be a universal Absolute Clock for measuring Absolute Time, even though no one knew how to construct such a Clock, even in principle. Let us examine this definition.

What is usually meant by the statement, "I was cracking a lobster in a restaurant in New York at the same time you were eating a ham sandwich in an airplane flying over Los Angeles," is that if at the moment of these particular actions we had both glanced at our local clocks we should have read the same time (disregarding the effect of time zones, etc.).

However, there are two additional implications: (1) Our clocks should have been previously synchronized, and (2) after synchronization they must both have run at the same rate, so that one never got ahead of the other (i.e., showed a later time). But how do we synchronize, and what does it mean to be "ahead"? If we are looking at a clock so far away that its light reaches us late, the "ahead" clock may appear to be "behind."

This poses a problem. Even if we managed to construct identical clocks, we should still have to allow for communication delays resulting in out-of-date information. The trouble is that there is no means of *instantaneous* transmission. Light (or radio) signals are the fastest thing we have, and even they introduce a delay at great distances. It is the *separation* of the clocks which is the source of the difficulty. For example, it is especially unclear what is meant by "simultaneous" if I happen to be cracking my lobster here on earth while you, an alien creature, are eating your sandwich in a planetary system of the Great Nebula in Andromeda some two million light years away. (A light year is the *distance* light travels in one year.) On the other hand, if we are sitting at the same table under the same clock in the same restaurant of the same city on the same planet, we know quite unambiguously what is meant by "simultaneous."

The requirement therefore is to find a definition which works as well for events which are separated in space. We shall adopt the convention introduced by Einstein for coping with this problem: Two events are said to be simultaneous if an observer stationed *midway* between them receives at the same instant the light signals sent out by the two events. This definition does not require two different clocks to establish simultaneity, and it references the two events to only one location, that of the midpoint observer. A reversed version of the process also affords a means of synchronizing clocks: the central observer sends out a signal, and clocks at the two endpoints are synchronized (set to read the same time) when they receive the message. Figure 7–1 illustrates an experiment

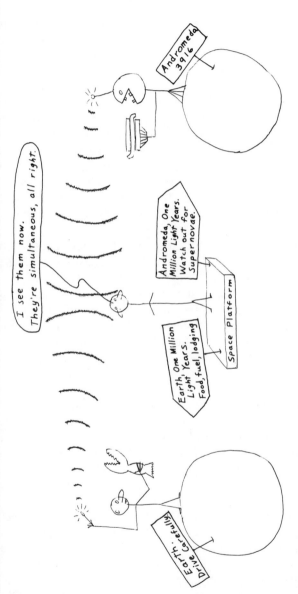

Fig. 7-1 *Definition of simultaneity.*

which defines simultaneity. In this picture two million light years of distance (assuming this is the distance between the planets) has been collapsed into the space of one page. Since either event is one million light years from the central observer, it will take him one million years to get the message. But this should not disturb us; in a *Gedankenexperiment* we do not concern ourselves with technological details, such as the requirement also that each diner at the appropriate moment turn on a light signal powerful enough to reach the observer, who must be looking in both directions at once. It is only important that we understand what we mean *in principle*. Actually, the observation station can be fully automated, replacing the two-faced observer with an electronic device known as a coincidence gate. The latter is triggered by the simultaneous arrival of light signals from the two directions, whereas it fails to respond to a signal from one direction only.

Now Einstein's is not the only way to define simultaneity, and on the face of it appears not to be terribly profound. But it has the advantage that it does not rely upon the agreement of different clocks, depending instead only upon the carefully measured equal distances from both events to the central observer, and the known property of the speed of light as a universal constant. The important thing is to have a consistent definition. We shall, however, discover that even such a careful definition leads us to some strange conclusions.

First we show that two events which are simultaneous for one observer are not necessarily simultaneous for another. This would perhaps come as no surprise if the observers were using a system depending on some cheap five-dollar watches. But even our precise definition of simultaneity still produces a "contradiction."

To illustrate this relative character of simultaneity, Fig. 7–2 shows a space platform similar to that in the previous figure. Now, however, we are concerned with events of a much more local nature. There are two observers, one standing on the platform, and another in a rocket ship passing it at constant velocity. The ship passes so close to the platform that for an instant they can be considered to be at the same position. An unusual accident occurs, and we examine the reports filed by the two observers, each using his own frame of reference. (Figure 7–2 has adopted the frame of the Platform Observer.) A bit of poetic license in reporting various time delays will serve to distinguish effects which are usually overlooked because of the relatively high speed with which light travels.

Platform Observer's Story

"At precisely the moment the Rocket Observer passed me, two meteors struck both ends of my platform simultaneously. (See Fig. 7–2a.) They left good-sized holes, and because the rocket ship was passing so close, tore holes in *its* front and rear as well. (Those rocket drivers are always trying to see how close they can come.) I know the meteors struck simultaneously, because I happened to be located precisely midway between the two points of damage on my platform, and the flashes from the two events reached me at the same moment. It's a good thing the insurance companies pay double for these simultaneous accidents.

"Unfortunately, however, the Rocket Observer's report is going to contradict mine, and I'll probably have trouble collecting. (The companies are always looking for technicalities to avoid making payments.) Since the Rocket Observer was passing right by me at the moment of the accident, he also was precisely midway between the two meteor events, so he will think he also qualifies as a judge of simultaneity. But the rocket ship and its entire frame of reference was moving ahead so fast he was able to meet the light wave emitted by the event up front before it reached me. (See Fig. 7–2b.) Since his report on the front meteor strike is earlier than mine, they'll probably think him more reliable. Actually, by the time light signals from both events reached me (Fig. 7–2c), he was still completely unaware of the fact that a piece of his tail had been blown off. It was only later that the signal from the rear event finally caught up with him. Even though he was midway when the meteors struck, he was no longer midway when the light signals reached him. So now he will say the front was hit before the rear, and all this confusion comes from his using a moving frame of reference. They warned us about that in space school."

Rocket Observer's Story

"I'm relaxing in my contour chair, when suddenly this platform whips past me at a tremendous speed, going backward. Shortly

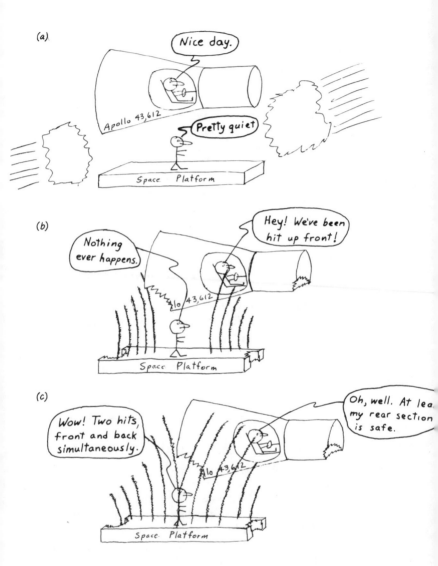

Fig. 7-2 *The relative character of simultaneity (The Platform Observer's story). (a) Meteors strike front and rear. But no one has discovered it yet. (b) Rocket Observer gets the message from front. But Platform Observer knows nothing, since light waves have not reached him yet. (c) Platform Observer gets the message from both front and rear. Rocket Observer still knows only about front damage, since waves from rear have not reached him yet.*

afterward there's a big flash up front, apparently a meteor strike, splattering holes in both my ship and his platform. Before I even have time to recover, there's another flash behind me; believe it or not, a second meteor putting a hole in the tail as well! Then the Platform Observer radios me this cock-and-bull story about both events occurring simultaneously. I know the front was hit first, because my chair happens to be precisely midway between the hole up front and the one in the rear of my ship. Those holes prove I never left the middle position, so I'm a judge of simultaneity. If the accidents had occurred simultaneously, both flashes would have reached me together—which they did not.

"I finally figured out what confused the Platform Observer. It was the backward motion of his frame of reference. He was rushing *away* from the earlier forward accident, and its light signal was therefore late in catching up with him. On the other hand, since he was moving *toward* the later occurring rear accident, he caught up with its light waves *ahead* of time. This made the earlier strike appear later to him and the later strike earlier, so that it seemed as if both events had occurred simultaneously. That's exactly the sort of thing they warned us about when we were studying moving frames of reference."

Which observer is "right"? Who is telling "the truth"? Were the events really simultaneous? The answer is that *both* observers are right, *both* are telling the truth, and the events are simultaneous for one observer and not simultaneous for the other. The only mistake being made by each observer is in thinking that his frame of reference is somehow "preferred" for seeing things as they "really" are, whereas the other frame is not. In other words, to the Platform Observer the platform is "at rest" in space, and the light signals come from the two holes in the platform. It is these holes, after all, which tell us where the meteors struck in the platform frame of reference. This is the view represented in the picture; the Rocket Observer's "mistaken" conclusion that there was no simultaneity in the making of these holes is due to his *motion* in this reference frame. But to the Rocket Observer the *platform* is moving (backward), and it is the *rocket* which is "at rest"; a picture (not shown) of these same events in the Rocket Observer's reference frame would therefore have the light signal emitted first from the forward hole in the *rocket,* and later from the rear hole; the Platform Observer's "mistaken" conclusion that the holes were made simultaneously is due to his *motion* in this reference frame. Either point of view is perfectly viable, and either observer can offer an acceptable explanation not only of his own observations, but of those of the other observer as well. Each reference frame is equally valid for pinpointing where the meteors actually struck. Light signals have the strange property (discovered in our Michelson–Morley experiment) that they travel with respect to the platform and also with respect to the rocket at the same velocity, despite the relative motion between rocket and platform. This is the apparent paradox. Since the Michelson–Morley experiment and the resulting theory of relativity have made us aware of a reality which violates our intuition with respect to the rules of simple addition, it is not surprising to find also that it leads us to the conclusion that a time relation can be both "true" and "not true," i.e., that two events can be both simultaneous and not simultaneous, depending only on the frame of reference from which they are observed.

Poet The redeeming quality I have heard ascribed to science is its ability to arrive at the objective truth—the fact that two scientists working independently must always eventually come to the same conclusion. This is what makes science beautiful. You are the Bearers of Truth, the messengers who speak directly to God. Now we hear that there are many truths. Are there also many gods? Perhaps there are many theories of relativity as well.

Scientist Your new-found disillusionment is with the prevailing mystique about physics, not with physics itself. You are beginning to discover the shallowness of the simplification which dismisses physics by calling it an "exact science," and assumes that the gift of subjective vision is the exclusive province of the poet. Science does begin with the assumption (and it is actually an assumption) that there exists an objective universe, independent of poets and scientists. But in our observation and interpretation of it we always introduce something of ourselves. For example, we have good reason to believe that hydrogen has precisely the same physical and chemical properties in another galaxy as it does here on earth. Does this mean that a scientist on another planet must necessarily arrive at the same *model* of a hydrogen atom as an earth scientist, or that he will be led to the same *Schrödinger equation* to predict its behavior? No, it does not. It means only that on both planets hydrogen and oxygen brought together under certain conditions will combine to produce water, and on both planets this water will boil at the same temperature and pressure. However, explanations by the two scientists of why this is so could turn out to be quite different. It is the experimental results their theories predict which must be the same.

P   But the fact that two events are or are not simultaneous *is* an experimental result, just like the temperature at which water boils.

S   Indeed it is. But I may wish to call the boiling temperature of water 100 degrees Centigrade, you may call it 212 degrees Fahrenheit, and a scientist on another planet may find it convenient to describe the same phenomenon without even introducing the concept of temperature. It is only after we agree on our definitions that we should expect to get the same result. In other words, only when we ask the same question do we get the same answer.

P   What do the rocket and platform observers have to do to get the same answer?

S   First of course they must decide what they mean by "simultaneous." (This presumably they have already done.) Then they have to agree on a frame of reference from which to observe simultaneity. If they do this, there will be no ambiguity.

P   Does this mean the rocket observer must leave his rocket and climb aboard the platform?

S   No. It means only that the rocket observer must be willing to accept the labeling of space (and time) by meter sticks (and clocks) which are *at rest* with respect to the platform; or alternately the platform observer must accept the labeling by meter sticks (and clocks) which are *at rest* with respect to the rocket; or they may compromise by accepting a frame of reference which is tied to neither the rocket nor the platform. But unless they can agree on a common frame of reference for judging such events, there is no way to designate *where* (or *when*) the two meteors struck, and the observers will argue endlessly.

P   It sounds rather like a game or legal argument in which the rules are made up as you go along. Apparently they turn out to be consistent rules in the end. And I can appreciate that the game has been amusing. But it is the fact that we all ended up as pieces on the game board which makes me wish you had removed the proceedings to some other corner of the universe.

S   Is it only the inventions which come from science that you fear, or is it the processes of reason and logic themselves?

P   I find that I can derive little comfort from these fine distinctions between science and the world which is the result of science.

questions for the reader

1. Suppose the meteors had struck in such a way that the Rocket Observer discovered them to be simultaneous. What would the Platform Observer have said in this case?

2. If you were the only object in the universe, what would be the meaning of the questions, "Where am I?" and "What time is it?"

3. Define the following: Constant velocity, loyalty, progress, stationary, patriotism.

4. Can you devise an experiment to define simultaneity independent of any frame of reference?

5. Would there be laws of nature if there were no people?

8   Space contraction and time dilation: nature's little jokes

Forgive, O Lord, my little jokes on Thee
And I'll forgive Thy great big one on me.

ROBERT FROST
In the Clearing

If the reader understands the relative character of simultaneity, he has taken an important conceptual step. If he is still grappling with the problem, he is in good company. It took the human race thousands of years just to discover the question.

We now investigate the consequences. Our *Gedankenexperiment* employs a pair of rods of a precise length, say ten meters. They may be of metal, or glass, or plastic; it does not matter. We may assume that both were produced in a factory which makes countless such identical rods for distribution to scientists, poets, space travelers, and platform observers all over the universe. Since these are difficult times, and neuroses and alienation are rife throughout the land, it was recommended at a psychiatric conference that all travelers carry such identical rods to promote mutual understanding and reassurance. "No matter what other disagreements they may have," the psychiatrists are reported to have said, "intelligent creatures everywhere can always begin a discussion by agreeing on the length of their rods."

Thus it is not surprising that the Platform Observer and the Rocket Observer, realizing they have reached an impasse on the issue of simultaneity of the meteor accidents, decide to cool it a bit by invoking this rod comparison test. The rocket and the platform pass each other once again at constant velocity, and as they do the two observers proceed to compare rod lengths. Things being what they are, it happens that this is a very fast fly-by; in fact, we will let the rocket pass the platform (and vice versa) at something like one-half the speed of light.

Now it is no small matter to compare two rods which are passing each other at such great speed. The procedure is as follows: The rocket rod is extended by the R.O. along the length of his ship fore and aft, and the P.O. orients the platform rod in a parallel direction. If the rods are of equal length, one pair of ends should pass each other at precisely the same instant as the opposite pair of ends. This may be seen in Fig. 8–1a, where the P.O. has stationed two assistants at the ends of the platform rod to inform him when these ends line up with the corresponding ends of the rocket rod as the rod passes. The P.O. is of course precisely midway between the position of his two assistants, who send him a pulse of light (or a radio signal) at the appropriate instant. The R.O. does the corresponding thing in his reference frame.

Assuming all goes well, each system should find the other's rod to have the same length as his own, and a consensus will have been obtained. But, as the reader may suspect, we are in trouble. What actually happens? One way to find out is to do the experiment. But this would be very expensive. We therefore *deduce* the answer to the question.

There are several possibilities to consider. One is that the P.O. finds the ends lining up simultaneously, thus concluding that the rods are of equal length, and that we live in the best of all classical worlds. This is the happy state of affairs depicted in Fig. 8–1a. But then what must the R.O. find? We know that if two events (i.e., the lining up of the two ends of the rods) are simultaneous in one frame they are *not* simultaneous in another frame moving at constant velocity with respect to the first. We discovered this in the last chapter. We also know from his observation of the meteors that in such a case the R.O. will have to report the coincidence at the forward end of his ship (right end in the picture) as having occurred *earlier* than the rear coincidence (left end in the picture), just as he observed the front meteor strike to be earlier. This is shown in Fig. 8–1b, where his delayed observance of the rear coincidence necessarily leads him to conclude that the platform rod is *shorter* than the rocket rod. The picture is one the R.O. would draw in describing what happened in such a case.

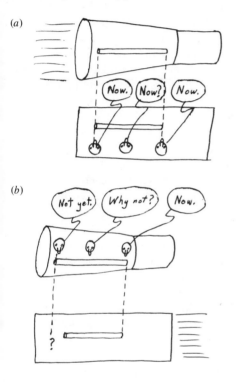

Fig. 8–1 Observers employing different frames of reference disagree on length measurements. (a) If in the Platform frame of reference the forward coincidence (lining up of front ends) is simultaneous *with the rear coincidence (lining up of rear ends), ... (b) ... then in the Rocket frame of reference the forward coincidence occurs* earlier *than the rear coincidence.*

Yet another possibility is that the R.O. finds the ends lining up simultaneously, in which case by precisely the same reasoning it will be the P.O. who says the other's rod is shorter. Thus regardless of which observer we assume tells us what we should like to hear, namely, that the rods are of equal length, the other observer always finds the "moving" rod shorter than his "stationary" rod. (Remem-

ber that in the rocket frame of reference the rocket is at rest, and
the platform is moving.)

Hence we see that it is simply not possible for *both* observers
to find the rods to be of equal length. Furthermore, since these
rods were produced identically in the same factory, and since neither
frame of reference for viewing them is preferred over the other,
symmetry considerations would make it impossible for them to be
of equal length for one of the observers and unequal for the other.
The only acceptable conclusion is that *both* observers find them to
be unequal. In fact, the relative motion between the two frames
must cause each observer to find the other's rod *shorter* than his
own. In other words, a rod in motion is shorter than the same rod
when it is at rest! This is true for either observer.

But then how can we explain Fig. 8–1a, which shows one observer finding the rods to be of equal length? We now know that
this picture corresponds not to the case of the two identical rods
we started out with, but rather to the state of affairs if a factory error
had caused the platform rod to come off the assembly line somewhat
shorter than the rocket rod. The foreshortening effect for moving
rods would then tend to compensate for the factory's error when the
P.O. observed the longer rocket rod, but would aggravate the discrepancy when the R.O. observed the shorter platform rod. Since
actually the rods were made of *equal* length, each observer must
find the other's rod shorter than his own. In any case, psychiatrists
to the contrary notwithstanding, the observers will surely fail to
reach an agreement.

All this "linear" reasoning has been rather strenuous. But this
is a relatively minor inconvenience when one considers the saving
in tax dollars we have effected by doing the experiment in our heads
instead of in space. The shortening of lengths in the direction of
relative motion, which we have just discovered in a qualitative way,
may also be derived quantitatively (see Appendix), giving us the
precise relationship between the length L of a rod moving with
velocity v, and its length L_0 when it is at rest with respect to the
observer; we merely state the result,

$$L = L_0 \sqrt{1 - v^2/c^2} , \tag{8–1}$$

where c is the speed of light, appearing in the equation because of
the way we have defined simultaneity.

It may be seen from this equation that the length L of the moving rod is always less than L_0, the rest length (except of course when $v = 0$). Since we have decided to have the rocket and platform pass each other at one-half the speed of light, in our case $v/c = \frac{1}{2}$, and the equation tells us that each observer finds the other's rod to be shortened from 10 meters to 8.7 meters because of its motion; his own rod, on the other hand, retains its original factory dimension of 10 meters, because it is *not* in motion in his frame of reference. As a matter of fact, he can use this same rod, which is at rest in his frame, for all length measurements. It is his meter stick.

Now if we permit the velocity v in the equation above to increase, the quantity inside the square root sign becomes smaller until, as we approach the speed of light ($v/c = 1$), the rod "disappears" altogether. If we were to continue the process beyond this, velocities *greater* than the speed of light would require us to take the square root of a negative number, which cannot possibly correspond to a real physical observable like length. Thus it is not possible for an object to attain a velocity greater than this universal speed of light with respect to an inertial frame of reference.

The shortening of lengths in the direction of motion is known as the *Lorentz–Fitzgerald contraction,* since it was originally proposed as an *ad hoc* hypothesis to explain away the results of the Michelson–Morley experiment. The suggestion was that there was an ether wind after all, and that our failure to observe it was caused by a mysterious shrinking of the Michelson–Morley apparatus by precisely this amount in the direction of the earth's motion through space. We shall have more to say about such hypotheses in a later chapter.

We have arrived at the Lorentz–Fitzgerald contraction merely by postulating the equivalence of inertial frames of reference, and by defining our terms carefully at every stage of the reasoning process. Clearly, it is the effect of the motion relative to an observer which is making the other fellow's rod appear shorter. *Appear* shorter? Or is it really shorter? One thing we know is that it will always come out shorter in any real or conceptual experiment. Therefore it is shorter. What else is there to say?

Poet You agree to the existence of an objective universe. In such a universe a rod is either shorter than another rod, or it is not. You can't have it both ways. In fact, you are apparently claiming that it is both shorter *and* longer, depending on who makes the comparison. It sounds as if what you have been describing is an optical illusion, not the way things are.

Scientist   This is no optical illusion.   A rod exists, and is observed by all kinds of creatures, or by no one.   But the word "length" has only that meaning which we choose to give it.   The objective universe exists (presumably), but it is *we* who have introduced the concept of length, and we have defined it in terms of a measurement (actual or conceptual).   The measurement requires some sort of meter stick, and this too we must provide.   Furthermore, without a definition of simultaneity we have no definition of length.

Let us go on with our story.   When the P.O. and the R.O. were issued the two identical rods, they were also provided with identical clocks.   These clocks were designed by the most expert of watchmakers, and were tested carefully in the factory over a period of years, during which time neither clock gained or lost the smallest perceptible fraction of a second over the other.   After the failure of the rod comparison test, in a final desperate effort at reconciliation the two observers decide to seek agreement with respect to times read on these clocks.   Just to make sure that nothing can go wrong, at the moment the rocket and platform pass each other they synchronize the clocks, so that both read precisely the same time, say three o'clock.   An hour later, when they are far apart, they compare them again.   Each observer has a very powerful telescope which enables him to see the other's clock even at the great distance which by now separates them.   But they must be careful about how the comparison is made.

"The observer must never forget," says the Observer's Manual (published *circa* 2070 A.D.), "that the clock comparison test requires a correction for the time lost in transmission. Thus if another observer has for the past hour been traveling away from your position at one-half the speed of light, it will take light then leaving his clock one-half hour to reach you; therefore if he left your position at three o'clock, check your clock again at four, but wait until four-thirty before looking through the telescope at his clock. Only then will you have a valid comparison."

We quote from this manual just in case the reader ever finds himself in a similar situation. It is only a detail of procedure, and as long as it is understood need not be considered further. The P.O. and the R.O. have been thoroughly schooled in these matters. (It was only the course in relativity theory which gave them some difficulty.) Thus they compare clocks correctly, taking account of time delays. And what do they discover?

If the P.O. should find that both clocks read four o'clock simultaneously (after proper correction), then the R.O. would find that they do not! Two distant events which are simultaneous in one frame of reference are not simultaneous in another. What are the two events in this case? Event A is the positioning of the hands of the P.O.'s clock at four o'clock, and event B is the positioning of the R.O.'s clock likewise at four. (The existence of a "neutral" midpoint observer may be imagined, even if he does not exist.) We know from our analysis of the simultaneity and rod comparison tests that if in the platform frame these events are simultaneous, then in the rocket frame the "forward" event (the positioning of the R.O.'s clock) occurs first, i.e., the rocket clock reaches four o'clock sooner. This means the R.O. finds his own clock to be running faster than the other; or, putting it the other way, he observes the platform clock to be *running slow*. The effect is illustrated in Fig. 8–2.

And by the same token, if the R.O. should find that the clocks read four o'clock simultaneously, the P.O. would find that they do not, declaring the rocket clock to be running slow. Hence we may conclude (using the same symmetry argument as formerly) that what actually happens is that *each* observer looking at the other's clock finds it to be running slow compared to his own.

This conclusion is derived quantitatively in the Appendix: Suppose the time interval between two events, say two ticks of a clock tied to the rocket (and therefore at rest in the rocket frame), is observed by the R.O. to be T_0. It is then found that the corresponding time interval observed by the P.O. using the *platform frame* (with respect to which the clock is moving) is instead

$$T = \frac{T_0}{\sqrt{1 - v^2/c^2}} \,. \tag{8–2}$$

Here v = velocity of rocket frame (and clock) with respect to the platform. A clock always runs slow when observed from a frame in which it is moving, and the reader can see that such "time dilation" is of precisely the same magnitude as the length contraction we discovered previously. This is of course no coincidence, since both effects are merely manifestations of the relative character of simultaneity.

Fig. 8–2 Observers employing different frames of reference disagree on time measurements. (a) If in the Platform frame of reference the events should be simultaneous, i.e., rocket clock and platform clock read 4 o'clock simultaneously, . . . (b) . . . then in the Rocket frame of reference the "forward" event will occur earlier, i.e., the rocket clock will reach 4 o'clock sooner.

There are many different types of clocks—mechanical clocks, electrical clocks, atomic clocks, biological clocks. There is nothing we have done in our *Gedankenexperiment* which stipulates the nature of the clock used by the observer, or the material of the rod in his length comparison. Time dilation and space contraction should therefore properly be considered properties not of the rods and clocks themselves, but rather of the space-time which they (and we) all occupy. Thus it is not only the other observer's clock which must run slow, but the other observer himself as well! His body processes after all constitute a form of clock, in this case a biological clock.

Poet And so the game continues. Black is white, left is right, forward is backward, long is short, late is early, old is young. It is as if, having become bored with your world, you have decided to replace it with one more entertaining. But I find it difficult to believe that hard-headed scientists (Stephen Spender calls you the "midwives of technological progress") would engage in such apparent sophistry without more evidence that this is the way things really are.

Scientist You are quite right. However, the story is told about Einstein that when reporters called to inform him that observations of a solar eclipse had proved to be consistent with the predictions of the general theory of relativity, he said, "I should be surprised if they had not." Scientists have a high regard for the process of reasoning; but all theories nevertheless remain suspect until their predictions have been independently confirmed by observations.

Our story about special relativity began with an experiment, and it must likewise end with one. Although accurate experiments to test the findings of general relativity, involving gravitational fields and frames of reference with changing velocities, were for a long time difficult to carry out, the predictions of special relativity have been relatively easy to confirm. In fact, the design of high energy particle accelerators would not be possible without taking into account some of the effects we have just deduced.

One of the most dramatic examples of time dilation is the decay of the *mu*-meson (designated by the Greek letter μ). This is a particle which is produced in the upper reaches of the atmosphere by the collision of cosmic rays from outer space with air molecules. We can also produce the same particle by artificial collisions in the laboratory, and find it to have an average life of some two microseconds (two millionths of a second). Of course just as the insurance companies have no way of knowing which of their customers will die first, so there is no way for us to predict how long any particular

μ-meson will live before suddenly decaying into an electron and a pair of neutrinos. (Actually, a meson is even more unpredictable than a person.) But we do know the *average* life expectancy of any given group of μ-mesons.

When a balloon observer (or a mountain climber) at an elevated altitude (and this time it is a *real* observer, not an imaginary one) counts the number of μ-mesons impinging on a detector he has carried up with him, and then makes the same measurement at lower altitude (or at sea level), he naturally finds fewer mesons below. This is due to the fact that few if any mesons are created in the lower atmosphere, which the primary cosmic rays cannot penetrate very deeply. As mesons created in the upper collisions descend, they proceed to die according to the appropriate actuarial statistics for μ-mesons. Since we are completely familiar with their behavior as a result of our laboratory experience, we are able to compare their survival rate with the corresponding survival rate of domesticated mesons in the laboratory. We can also measure their speed of descent, which turns out to be very close to the speed of light ($v/c = 0.992$). And when we do all this we find that the atmospheric mesons live significantly longer than the laboratory mesons. In fact, instead of the normal two microseconds lifetime allotted mesons here on earth, their atmospheric cousins have an average life of more like 20 microseconds. Furthermore, this increase is just that predicted by the time dilation equation (8–2) obtained as the result of our *Gedankenexperiment!*

The significance of all this is pointed up in Fig. 8–3. In the balloon observer's frame of reference a clock tied to the balloon runs at the "proper" rate. But when he observes a clock riding with the μ-meson (in this case the meson's lifetime is itself the clock), he finds the meson clock to be running slow. If he knows nothing about relativity, he will find this most puzzling.

The reason the experiment is so impressive is that the meson travels at a velocity close to the speed of light; hence the dilation effect is large enough to be unmistakable. Of course if the mesons were smart enough to do the same experiment on *us*, then in the meson frame of reference the balloon observer and all the rest of the world would be falling *upward* at this same very great speed. And the mesons would discover that instead of living somewhat less than a hundred years (like the "artificial" people created in their meson laboratories), human beings on earth actually live close to a thousand years (as observed by mesons, of course).

Fig. 8-3 *Decay of the μ-meson.*

Poet Does this suggest a solution to problems in geriatrics and longevity? If you could somehow get the earth moving at the speed of the μ-mesons, would we all live a thousand years?

Scientist Unfortunately, this question has no meaning. We are *already* traveling at this high speed in someone's frame of reference, namely, that of the mesons. It does not help to extend one's life in someone else's frame of reference, since one's own clocks (including one's body) remain at rest in one's own frame, and therefore proceed at the "proper" rate when observed by oneself. So even if we could persuade Congress to appropriate the necessary funds, people would still feel they were being cheated, since they would never notice the difference. I'm afraid that if you are to derive any benefit from this knowledge, it will have to take some other form than the strictly technological, at least for the present.

On the other hand, time dilation would certainly affect our lives if *some* of us could be brought to such high velocities relative to *others*. This is the so-called twin paradox. A pair of twins is separated early in life; one leaves the earth on a long trip in a rocket moving at a constant velocity close to that of light, while the other remains behind. At a prescribed place and time he turns the rocket around quickly and returns home at the original speed. When they are reunited the earth twin discovers the effect of time dilation; his brother has failed to age as rapidly as people on earth. All the clocks in the space ship, whether mechanical, electrical, or biological, have been running slow because of their high speed with respect to the earth. Hence the space traveler is still young, while his brother is now an old man. (In Fig. 8–4 we have placed tradition above timeliness in giving the old man the beard instead of the young one.)

This is not just a joke. It is the counterpart of the meson experiment. Although we have been unable to attain such high relative velocities between human beings, we achieve them every day between people and particles in the laboratory. In the workaday world of the high energy accelerators relativity is just business as usual.

Poet Something disturbs me about your twin paradox. You have gone to great lengths to make the point that velocity is relative.

If I take you at your word, then I must conclude that in the rocket's frame of reference it is the *earth* which is traveling at high speed; therefore from the viewpoint of the rocket twin it is the earth twin who should remain young. I suppose it is just as "logical" for you to claim that each one is younger than the other as it is that each one's rod is shorter than the other's.

Fig. 8-4 *The twin paradox. Space twin leaves for a visit to a distant planet, . . . and returns to find a generation gap.*

It is indeed appropriate to invoke the symmetry argument. As long as the rocket maintains a constant velocity, everything will remain symmetric. Each twin observes the other through a telescope (of course making the appropriate corrections for time of light transmission), each one concludes the other is aging less rapidly, each one sends messages claiming that his brother has the better deal. However, this condition prevails only so long as both earth and rocket remain inertial frames of reference, i.e., only so long as both maintain a constant velocity.

If the twins are ever to be reunited, a day of reckoning must come. On that day the space twin will fire up the rockets which reverse his direction and send him back to earth. It is this *change* of velocity which breaks the symmetry. If the space twin happens at that moment to be eating a bowl of soup, it will spill in his lap. He will experience the familar "g-forces" all astronauts encounter during rocket thrust. In fact, the space twin is rather likely to be crushed against the bulkhead of his ship, if we require that he make a quick turnaround followed by a smooth and uniform homeward trajectory. (This turnaround problem happens to be one of the serious practical obstacles to the expedition.)

On the other hand, the twin back on earth will remain completely unruffled, and can finish a bowl of soup while his brother's eyeballs are popping. This is what breaks the symmetry. Unlike the earth, which we may consider essentially an inertial frame, the rocket ship does *not* remain an inertial frame throughout the journey; hence we may not employ this line of reasoning for predicting results from the viewpoint of the space traveler. Only the earth twin can use the principles of special relativity in order to determine what will actually happen, since when we look at things from his point of view we are always using an inertial frame of reference.

Poet But, except for the brief turnaround time, the rocket frame does remain inertial for most of the journey. Therefore most of the time the space twin should observe his brother's clock to be running slow, keeping the earth twin young. How can you reconcile this with your claim that when the traveler returns, he will find his earth brother to be the *older* of the two?

Our time dilation formula may be applied only from the viewpoint of the earth twin, since the earth is the inertial frame of reference. A telescope on earth simply reveals that the space traveler consistently runs slow throughout the journey, and hence retains his youth. However, it is also true that the space twin during the constant velocity journey outward (and also during the return) finds the *earth* clock to be running slow, and his brother aging less rapidly than himself. There is only one explanation which can reconcile the apparent contradiction between this requirement and the asymmetric condition of the twins at the end of the trip. Something must happen *during the turnaround*. A detailed analysis of this critical period from the viewpoint of the non-inertial rocket frame of reference would involve the theory of general relativity, to which the reader has not been introduced. But he does not need to know general relativity to know the answer, since there is only one possible conclusion which permits their ages to come out different at the end of the trip. If the space twin could somehow manage to observe his earth twin's behavior while the space ship is reversing its direction, he would be shocked to discover that this young and vigorous brother back on earth is for no apparent reason suddenly turning into an old man! Thus it should come as no additional surprise to him when he finally arrives home to discover that this has indeed happened.

Poet How did all this lead the world into atomic energy and bombs?

Scientist I want you to notice that you too are unable to confine your curiosity. You do not draw any lines; you want to know everything, just like the scientists. We were discussing fundamental principles. How did we get on the subject of bombs?

questions for the reader

As a result of consistent failure to agree on anything, there has been a deterioration of relations between the Platform People and the Rocket People. An arms race has begun, and each side is convinced that the others are out to destroy them in order to prevent them from destroying the others first. At three o'clock, when the

Platform People are holding a Red Alert to test the effectiveness of their defenses, the Rocket passes them at a constant velocity equal to one-half the speed of light. As is the custom on such occasions, everyone synchronizes clocks. However, to the surprise of the Platform People, the Rocket People are also holding a Red Alert.

"It would be quite a coincidence," say the Platform Strategists, "for them to be holding their test Red Alert simultaneously with ours. What seems more likely is that they do not believe ours is only a test alert, and they have therefore started the real thing. If that is true, then according to our calculations they should be launching their first strike missiles in just one hour from now, at four o'clock. On the other hand, if it is only a routine alert, it will be all over by three-thirty and they will be returning the covers to their missile silos. We have instructed our surveillance outposts, whom the Rocket will be passing in half an hour, to keep us informed. If by three-thirty their countdown is still proceeding, we will launch our antimissile missiles."

"Hold on a minute," says the Platform Observer. "That is a pretty extreme measure. Suppose there is a miscalculation, and it is only a test alert after all."

"We have considered that possibility too," say the Platform Strategists. "The safest thing to do in such cases is to proceed on the basis of their capabilities, not their intentions. The beautiful thing about our antimissile missiles is that they are programmed to attack only other missiles. Civilian populations and cities have absolutely nothing to fear. If they fail to encounter any enemy missiles, they will change course before penetrating the Rocket People's radar screen, and no harm will be done. Our system is absolutely fail safe and fool proof. Those antimissile missiles secure the credibility of our nuclear deterrent; they are our guarantee against an enemy miscalculation. Of course if the Rocket People have really started a war, we will have the jump on them. We will see the flashes when our antimissiles destroy their strategic missiles, and with their first strike capability wiped out, we can really clobber them."

Unfortunately the Rocket People also have their Strategists, who read the same books as the Platform Strategists and have reached the same conclusions. They too suspect the Red Alert of the Platform People, and have embarked on precisely the same course of action.

1. Describe what happens at three-thirty in the Platform frame of reference.

2. Which of the following statements is most nearly correct?
   a) The Platform People launch their missiles before the Rocket People.
   b) The Rocket People launch their missiles before the Platform People.
   c) Both sides launch their missiles simultaneously.
   d) None of the above.

3. The Platform People and the Rocket People have completely destroyed each other's populations, except for one sole survivor on either side. These survivors are now arguing their respective cases before the Galactic Historian, in an effort to get the record straight. The subjects at issue are:
   a) Who was the agressor?
   b) Who won the war?

   The Platform Survivor argues, "Clearly we were not the aggressors, since the missiles we launched first were strictly defensive weapons, and if the Rocket People had been as peaceful as they claim, no harm would have been done. On the other hand since we got the jump on them, we must have won the war, although I admit things got rather confusing for a while."

   Compute the time the Platform People launched their antimissiles, and the time the Rocket People launched theirs, all in the Platform frame of reference. (You will need to use the time dilation equation (8-2) for determining time elapsed on a Rocket clock since clocks were synchronized, as observed in the Platform frame of reference.)

4. Now compute the times of these two events in the Rocket frame of reference. Thus you should discover that it is possible for an event A to precede an event B in one frame, and yet have B precede A in the other. This may perhaps disturb you, by raising questions about the relationship between cause and effect.

5.   Show, however, by considering the locations and times of these events in the Platform frame of reference, that once the missiles were launched by the Rocket People, information about this launching event (traveling even at maximum speed in the form of light or radio signals) could have reached the Platform People only *after* they had already launched *their* missiles. This is an illustration of the fact that when two events occur in reversed order in different frames of reference, neither event can actually influence the other. Thus there is no violation of the principle of causality, which links cause and effect. Otherwise a father could be born before his son in one frame, and the son before the father in another frame. Fortunately relativity does not impose such awesome conclusions upon us; the father always precedes his son no matter what the frame of reference. It can be shown that two events which are connected by a causal relation (one being the cause and the other the effect) will occur in the *same* time order in *all* frames of reference, despite the existence of time dilation. But when events are so far apart in space, or so close together in time, that light sent from event A arrives at location B only after event B has already occurred (and vice versa), only then can the order of the events be reversed in another reference frame. In such a case, A and B will have no influence on each other. This was the case with the two launchings. Both the Platform People and the Rocket People have only themselves to blame for what occurred, and the Galactic Historian will so record it.

9   Mass into energy: The human race discovers what makes the stars shine

The results of the preceding investigation lead to a very interesting conclusion, which is here to be deduced.

A. EINSTEIN
"Does the Inertia of a Body Depend upon Its Energy Content?" Annalen der Physik, 17, 1905

The paper quoted above followed immediately after Einstein's original work, "On the Electrodynamics of Moving Bodies," in which he defined simultaneity and deduced the relativity of time and space. In order to see how Einstein reached his conclusion, we shall have recourse to a slightly more quantitative development than has been our practice until now; the less intrepid reader who elects to omit part or all of this chapter will be forgiven.

Energy is important to a physicist. Thus one could expect that, having led the Rocket Observer and the Platform Observer to compare lengths and times, we should also have asked them to compare energies. But first we must have a definition of energy. We shall avoid precise mathematical formulations, and state rather generally that energy is the ability to do work. Of course it comes in many forms. In a tank full of fuel oil, or in an electric battery, it is chemical energy. Our house current carries electrical energy. In the case of water spilling over the top of a dam to drive a generator, it is called gravitational potential energy. A skier who clears an abrupt bump or rise in the ground is momentarily hurled into the air against the opposing downward force of gravity; this is made possible by his "kinetic energy" of motion, which is due to the speed with which his body mass approaches the rise. Every moving body has kinetic energy simply as a result of its motion, and the amount of such energy is proportional to the mass of the body. Now if you have had a course in high school physics, you may remember that kinetic energy T is given by

$$T = \tfrac{1}{2}mv^2, \tag{9-1}$$

where m is the mass of the body, and v is its velocity. It is this quantity which determines how high into the air the ski jump can actually "throw" the skier.

Light waves carry energy also; if you doubt this, watch the sunlight set the vanes of a toy radiometer into rotational motion. Actually, even though the light reaching us from the sun exerts a relatively gentle force upon us, it is the original source of virtually all the energy available to us on earth.

Solar energy passes through many forms. It is converted by plant life into chemical energy which eventually takes the form of the coal or oil used to power our industries (and pollute the air). It heats the water in lakes and oceans, causing it to evaporate, so that it can descend again as rain on the mountain tops. This water may then be dammed and used to drive generators before it makes its way back to the lakes and oceans once more.

Physicists have discovered that, regardless of how energy may be converted, it is neither created nor destroyed. It merely reappears in various forms. This observation, known as the principle of energy conservation, is a fundamental physical law. Of course laws are made by people, and may therefore be modified. But changing a law of physics is a serious matter, somewhat comparable to changing the Constitution of the United States. We change a law of physics when it no longer offers a reasonably simple explanation of the world around us. But it is not done frivolously. And we disturb as little of the existing structure as possible. If a new principle leads us to make repeated changes or additions in every physical law ever conceived, this constitutes an *ad hoc* sort of physics (becoming the subject of our next chapter), and we should probably examine the new principle rather critically before tossing out everything else. The structure of special relativity developed in the preceding chapters was based on the overturning of only one principle, the rule for the addition of velocities. And this modification was based on very substantial experimental evidence.

Now if in the days when the Platform and Rocket Observers were still able to conduct scientific experiments, we had asked them to compare energies, we have no reason to expect that they would have agreed here any more than they did on anything else. In fact, unlike length contraction and time dilation, which are strictly relativistic effects, a disagreement about energies should come as no surprise at all, even to the P.O. and the R.O. In the platform frame of reference, the P.O. says the R.O. is moving, and hence has kinetic energy of motion. But in the rocket frame of reference the R.O. is at rest, and observes that he himself has *no* kinetic energy; in fact it is the P.O. who now has the kinetic energy. All this is perfectly normal.

What would *not* be normal would be for an observer in one frame to find that energy is *conserved* with the passage of time, and an observer in another frame (moving at constant velocity with respect to the first), to find that it is not. This would not only violate the classical law of energy conservation; it would even violate our new first postulate of special relativity, which states that there is no way to single out one frame of reference over any other. If there were a frame in which energy is conserved, and another in which it is not, it would become possible for an observer to know whether his is the "preferred" frame. Thus the failure of energy conservation in one frame would destroy not only our old physics, but our new principle as well. Yet this is what happens, as Einstein showed in a *Gedankenexperiment*—unless we are to recognize the existence of a completely new source of energy.

Suppose the R.O. faces the P.O. as they pass each other, this time at ordinary twentieth century rocket velocity, much less than the speed of light. The P.O. stretches out his arms, holding a flashlight in either hand, as in Fig. 9–1. He turns on both flashlights for a prescribed time period, and then turns them off. The rocket is assumed to be passing so close that its passengers, as well as those on the platform, can observe all the light emitted by the flashlights. (This may not be readily apparent in the picture.) Now the total amount of light energy which flows from these two flashlights may be calculated. We shall not perform the computation; however, not only was Einstein able to do it, but even the P.O. and the R. O., who never went much beyond classical physics in school, are capable of doing so likewise. In any case, whether they compute it or actually

Fig. 9–1 *The transformation of mass into energy. More light energy is delivered by the two flashlights as observed in the "moving" rocket frame than in the platform frame which is at rest with respect to the flashlights.*

measure it, either observer can determine the energy flow in his frame of reference, and the results turn out to be *different* in the two frames of reference. This is related to the so-called *Doppler effect* which one discovers when the siren of a passing ambulance changes its pitch as a result of its motion with respect to the observer. A similar *Doppler shift* occurs in the color of starlight when a star is moving away from us, as the stars are apparently doing in our expanding universe. In the case of water waves an example of differing wave energies in two frames of reference is the fact that the moving water skier jounces rapidly up and down on the waves, while the "stationary" swimmer sees the same water rising and falling quite gently (Fig. 9–2). All these phenomena are manifestations of the fact that different observers measure different wave energies depending on their relative motion.

Fig. 9–2 *Wave energies in different frames of reference. From the viewpoint of the water skier, i.e., in a frame of reference in which he is at rest, the passing water waves have considerable energy, . . . But in the frame of reference of a swimmer floating in the same water, all is quiet and peaceful, as the waves lap gently by.*

Thus it is rather to be expected that the total light energy observed from the two flashlights is not the same for the two observers. The emitted energy in fact turns out to be greater as observed in the "moving" rocket frame of reference than it is in the "stationary" platform frame, in which the flashlights are at rest. Perhaps the reader will not even be particularly surprised to learn that the energy increase due to this motion turns out to be precisely the same square root factor we encountered in the contraction of lengths and the dilation of time. In other words, if E is the total energy emitted by the two flashlights in the stationary P.O. frame of reference, then it can be shown (see Appendix) that the R.O. in the moving frame will find this energy to be rather

$$E' = \frac{E}{\sqrt{1 - v^2/c^2}} , \qquad (9-2)$$

where v is the relative velocity of the rocket with respect to the platform, and c is, as formerly, the speed of light. (This time, however, the numerical difference betweeen E and E' is rather small, because we have decided to use a small relative velocity, so that v/c is much less than one.)

The question Einstein posed (and we refer it to the observers and to the reader as well) is where does this extra energy for the moving observer come from? We know that the light energy is derived from chemical sources in the flashlight battery. But a battery is a battery in anybody's frame of reference. It is manufactured with a prescribed amount of available energy. Once it runs down, it will no longer light the flashlight, no matter in which frame of reference you ask the question. If in the platform frame the energy E carried away by the light is equal to a given amount of chemical energy which was originally stored in the batteries, as one would expect, then in the rocket frame the greater light energy E' will clearly *not* be equal to this previously stored chemical energy. Thus if energy is *conserved* in one system, it is apparently *not conserved* in the other.

But we do not have sufficient grounds here for discarding the long-established principle of energy conservation. Let us rather look, therefore, for another explanation, something which permits energy to be conserved in *both* frames. Where can we find a source for the greater amount of light energy seen in the moving rocket frame? One thing we do know is that this added light energy is due somehow to the *motion* of the flashlights in the rocket frame of reference. Hence, Einstein deduced, its source ought to be a form of *kinetic* energy which is being depleted, according to an observer in the rocket frame. And kinetic energy depends on only two things, the mass of a body and its velocity. Thus if the lighting of the flashlights causes more energy to be lost by the batteries in the rocket frame of reference (so as to supply the extra light energy in this frame), then there are only two possible sources for this kinetic energy loss—mass loss and velocity loss. But in our *Gedanken-experiment* the relative velocity of the rocket (and therefore of its frame of reference) remains constant. Certainly the turning on of two flashlights in precisely opposite directions will not change it. Hence there remains only one explanation: *mass* is being lost.

In other words, the P.O. and his flashlights must suffer a reduction in mass when the flashlights emit light! (In this case of course it would be the batteries which experience the mass loss.) Now in the platform frame this loss of mass cannot change the kinetic energy of motion, since in this frame there is no motion; i.e., the velocity of the flashlight batteries is zero. But in the rocket frame of reference the P.O. and his flashlights are all moving to the left at velocity v, and therefore have kinetic energy due to such motion. Thus a loss of mass will produce in the rocket frame a loss of kinetic energy. And this energy loss in the rocket frame must be just enough to supply the additional light energy emitted in this frame.

We have designated the light energy in the stationary platform frame as E, and in the moving rocket frame as E'. The difference in energy is therefore

$$E' - E = \frac{E}{\sqrt{1 - v^2/c^2}} - E, \qquad (9\text{--}3)$$

from the expression (9–2) we have written earlier to relate to E' and E in the two frames of reference.

Einstein proceeded to set this energy difference equal to the loss of kinetic energy, $\frac{1}{2}mv^2$, where m is the amount of mass which is lost during the time the flashlights are on. In other words,

$$\frac{E}{\sqrt{1 - v^2/c^2}} - E = \frac{1}{2}mv^2. \qquad (9\text{--}4)$$

This expression involves E, m, v, and c, where E is the energy of the emitted light as seen in the "stationary" platform frame, and m is the mass lost by the batteries during the time the lights are on. It is possible by algebraic manipulation to cancel v; if we do so, and we simply state the result here, we obtain the famous expression

$$E = mc^2. \qquad (9\text{--}5)$$

(In doing the algebra a slight approximation is necessary, as shown in the Appendix, taking advantage of the fact that the v^2/c^2 is very small in our experiment compared to the number one. At "ordinary" rocket velocities, say 186 miles per second, $v^2/c^2 = 0.000001$, which is indeed a lot smaller than one. However, the mass-energy relation we have just obtained may actually be shown to be exact even for higher velocities, provided we are careful about how we define kinetic energy.)

Thus in order to retain our principle of conservation of energy, we have been forced to conclude that the energy emitted by the flashlights in the platform frame is equal to the mass lost by the batteries, multiplied by the speed of light squared. The battery energy, which we have until now simply referred to as chemical energy, is really due to a loss of mass in the material of which the batteries are composed. We now know that this loss occurs in the total mass of the electrons which surround the nucleus of every atom. Matter has actually been converted into energy, although the amount of energy loss in this case is considerably smaller than it is in processes where the nucleus itself loses mass, such as those which occur naturally in the interiors of stars, or are induced artificially in nuclear "explosions." It is interesting that we started out looking for an explanation of the *difference* in energies $(E' - E)$ observed in the two frames, and we have ended up with a discovery of the *source* of the energy E.

Poet   Did Einstein then appreciate what he had wrought?

Scientist   Not in the way you mean it. To him it was an important fact of life, but certainly not a weapon. He proposed testing his conclusion on radioactive materials where the loss of mass would be significant because of the large amount of energy radiated in such processes. The experiment was done years later, of course confirming his predictions. It was only after many decades, when more had been learned about the atomic nucleus, that it was realized chain reactions could be obtained, with enormous release of energy through the conversion of mass. Not only were weapons farthest from Einstein's mind when he wrote his paper; he did not even realize he had uncovered the secret of what makes the stars shine.

P   So that is how it began. We had been muddling along. A bloody war every so often, now and then a plague or a famine, or maybe a couple of earthquakes. But we always managed to recover. And then we really made it: 1905, the year for sowing the wind. Which year is for reaping the whirlwind, 1975? 1985? Or are we likely to make it to 2000?

S   You remind me of the mother whose adolescent son has just acquired a driver's license. Now she lies awake at night waiting for him to come home. Everything was all right before; why did he have to learn to drive?

P   There is an important difference. This adolescent has the entire human race in his car.

S   Well, there is only so much you can do with an analogy. Anyway, the problem is how do we get through our adolescence, not why can't we be a little boy again. It is possible that every intelligent race in the universe passes through a similar crisis, and that some manage to survive, while others do not, just like teenage drivers.

P   And what do you think of our chances?

S   Somebody estimated two per cent probability of all out nuclear war per year, as long as we continue to live the way we did before "learning to drive," with every nation looking out for number one. On this basis the probability of getting through the next decade is something like 82 per cent, and roughly fifty-fifty that we shall make it to the turn of the century. You can see why the young people are so edgy these days. Under ordinary circumstances we could expect them to wait patiently like good soldiers while they acquire the profound maturity and wisdom of their elders. But unfortunately there is no assurance that this will ever happen. Of course these probability estimates are not altogether meaningful, since actually we have no data on similar situations. Perhaps somewhere in the universe statistics are being compiled on what proportion of intelligent races survive their adolescence. If we manage to get through the crisis, some day in the distant future we may be privileged to see such data, and then we shall know what our chances were all along.

questions for the reader

1. Consider a vehicle powered by converting the mass of one kilogram (2.2 pounds) of nuclear fuel completely into energy. Determine how many gallons of gasoline would be required by a conventional engine in order to do the same job. If you express mass in kilograms, and take the velocity of light as 300,000,000 meters per second, the energy in Einstein's expres-

sion will come out in joules.   A gallon of gasoline delivers
approximately 100,000,000 joules when it is burned in the
conventional way.

2. Check the scientist's estimate of our chances of averting a
nuclear war if we continue to conduct business as usual, assuming a two per cent probability that such a war will be precipi-
tated in any particular year. (The way to do this is to note that
since the probability that it will *not* happen during the first
year equals 0.98, then the probability that it will not happen
during the first year and also not during the second year is
0.98 × 0.98, and so on.)

10   # On the nature of *ad hoc* theories

"Beauty is truth, truth beauty"—that is all
Ye know on earth, and all ye need to know.

JOHN KEATS
Ode on a Grecian Urn

We have come a long way. Is all that has been said just a matter of physics? What does it mean for the human race to have reached the stage where it is capable of this kind of reasoning? It is appropriate to interrupt our story in order to address ourselves for a while not to the physical world, but to the thought processes which enabled us to learn about it. This, after all, is what is most important. The reader is unlikely to make a major contribution to physical science, but in the way he thinks or tries to solve his problems he may have much to say and do about the world which he will continue to share with the physicists.

When the Russians orbited their first astronaut, Yuri Gagarin, an interesting interpretation of this unusual event was offered by a friend of mine. "It surprises me," he said, pointing a finger at the newspaper, "that it never occurs to anyone to question all those claims the Russians are making. After all, they are perfectly capable of fabricating such a story out of whole cloth."

"We have independent checks," I said.

"What checks?" he asked.

"Well, we can track the satellite by both optical and radio means."

"That doesn't prove there's a man up there," he said.

"But we pick up his radio signals," I pointed out. "We hear him talking."

"That's easy," he said. "Anybody can put a tape recorder in a space capsule and claim it's a person. He probably recorded all those messages before the launching."

"But he answers questions," I protested.

"So what?" he said. "It's no trick to synchronize prearranged questions with triggered responses. I don't think you appreciate the sophisticated devices modern technology makes available for perpetrating such a hoax."

My friend was not joking; he meant every word. Furthermore, he was absolutely right. I came to realize this as he proceeded to demolish one objection after another with completely plausible counterarguments. There was no way in the world to prove with certainty that there was actually a man up there. My friend was capable of expanding his *ad hoc* theory (that is what such a patch-work construction is called) more rapidly than I could devise or carry out experiments to break it down. If our space agency had launched a reconnaissance satellite and photographed Gagarin through the window of his capsule, then it could have been a screen and movie projector the Russians had installed with fiendish clever-ness, precisely in anticipation of such an eventuality. If we were told that astronauts had effected a rendezvous, boarded the vehicle, and shaken hands with him, then I am certain one could even invent an explanation for this to support the theory. After all, there are such things as robots, or the Russians could have infiltrated NASA with agents instructed to perpetuate the fraud.

In all fairness it should be pointed out that my friend is an intel-ligent and stable individual, with a fertile imagination and even a certain amount of scientific training. I am certain that by now he has abandoned his theory, although I have not questioned him lately. He is neither paranoid nor disoriented, and, like most of the human race, is normally capable of being dislodged from one of his favorite obsessions before things reach a dangerous pass. But somehow in the course of his education he had not been properly apprised of the pitfalls of the *ad hoc* argument.

Unfortunately the means for testing a scientific hypothesis are not always as straightforward as they were in the conjecture about the foreign astronaut. No one has ever seen a hydrogen atom in operation. Furthermore, as we shall discover in subsequent chapters, we have excellent reasons for believing that no one ever can or will. Therefore we have to be very devious in inventing experimental devices for tricking nature into revealing itself unto us. We cannot enter the nucleus of an atom in person; hence we find ourselves in the delicate position of the detective who has suspicions about a house he is watching, but cannot investigate it directly without destroying the evidence. He therefore resorts to minor provocations like throwing rocks through the window, or as an extreme measure perhaps even firing bullets through the walls, in the hope that such prodding will eventually force the house to reveal its contents. In physics this is known as the scattering experiment. Much of our knowledge of the submicroscopic world has been obtained by firing projectiles at poorly understood atomic or nuclear targets and then using the evidence thus obtained to reconstruct the "crime."

Now if our detective has adopted scientific methods of reasoning, he will not content himself with picking up everything he can find and simply hurling it at the house. Instead he will say, "There are the following possibilities. If hypothesis A is correct, a rock should produce these results. On the other hand, if hypothesis B is the correct one, then one should expect something else." Only then will he throw the rock. Afterward he will reexamine the hypotheses in the light of the evidence, try to come up with a modified theory consistent with the new data, and finally devise yet another experiment to test the new theory. But if he does not beware of *ad hoc* reasoning or an inordinate desire to prove his initial conjecture, he may end up with an explanation which, although "logical," will never stand up in a court of law.

The experiments to test the ether hypothesis were ingenious devices to verify what virtually all the scientists then believed. The fact that Michelson and Morley (as well as other experimentalists) were able to establish beyond a reasonable doubt that they had failed, was a triumph of reason over hope. There have been many attempts to salvage or revive the ether theory in an effort to "disprove" Einstein's simple though traumatic conclusion that one and one does not have to equal two. Some of these have been useful and constructive; others have been the work of cranks who never really bothered to understand the structure they were trying to demolish.

The reader may consider himself as yet poorly prepared to make a critical assessment of the findings of modern physics; but it is important for him to be able to recognize whether a theory propounded on any subject at all is more concerned with being "right" than with being simple.

We shall not explore all the alternatives to the theory of relativity. But it may be useful to experience the flavor of a typical argument. It could proceed as follows:

Ether Theorist I do not believe that light waves can be transmitted without a medium. Water waves employ a medium—the water. Sound waves have a medium—the air. Something has to be waving. Therefore electromagnetic waves (like light and radio signals) must also be transmitted via a medium. It is simply not plausible for them to be transmitted by nothing at all. It is likewise implausible that the velocity of wave propagation should not follow the normal rules of addition, or, as you have yourself put it, that one and one is not two.

Relativity Theorist The fact that certain other types of waves require a material medium is no proof that this must be true for *all* waves. When you invent a medium no one has ever observed, you are introducing an unnecessary complication. I admit that it was painful for me to give up the old rule for adding velocities. But this rule was based on everyday experience at low speeds. The new principle accounts for both the old experiments and the new ones. In any case, one's intuition cannot be trusted in a domain in which there is no previous experience. The theory of relativity is consistent with all the experimental evidence and has made some rather startling predictions which have been completely borne out.

ET But I can explain all these observations within the framework of the ether theory; and I do not have to give up anything as fundamental as the rule of addition.

RT How do you explain the failure of the Michelson–Morley apparatus to detect an ether wind resulting from the movement of the earth in its travels around the sun?

ET This can be explained by a shrinkage of their apparatus by just the right amount in the direction of the earth's motion. Your theory also has this contraction of lengths.

RT But I *deduce* this contraction from the simple assumption that the laws of physics are the same in all frames of reference, and the fact that the speed of light is the same. Otherwise why should the apparatus shrink?

ET I don't know. Why should the rule of addition be wrong?

RT All right. Let's accept space contraction as a postulate in your theory. But the experiment has been repeated with a modification in which the arms of Michelson's interferometer were of unequal lengths. And in this case it was not enough to introduce the Lorentz–Fitzgerald contraction. In order to preserve the ether theory you also have to invent time dilation. Now not only does the apparatus shrink, but clocks have to run slow in a moving frame of reference.

ET So what? Your theory has time dilation too.

RT Not as a postulate.

ET Nevertheless, I still have a right to believe in the ether.

RT Yes. You also have a right to believe in hobgoblins.

ET What is the difference between your postulates of special relativity and a theory of hobgoblins?

RT Simplicity.

ET At least my theory preserves the rule of addition.

RT What about the successful prediction of relativity that matter can be converted into energy?

ET My theory does not contradict this. And at least it preserves the rule of addition.

Thus we see that whatever the established principle we are committed to uphold at all costs, whether it be American motherhood and apple pie or the rule of addition, the *ad hoc* approach leads us to construct an endless chain of supporting logic in an effort to demolish unfavorable evidence. The above dialogue is not intended to prove the "correctness" of the theory of relativity, or the "incorrectness" of the ether theory. The latter can be made a perfectly consistent structure. The physicist rejects it primarily on esthetic grounds.

What makes the *ad hoc* theory scientifically unattractive? Its lack of simplicity. Paul A. M. Dirac deduced the intrinsic spin of the electron and predicted the existence of antiparticles by combining the principles of relativity and quantum mechanics in a masterpiece of pure reasoning, simplicity, and elegance. He once made the observation that if a decision is to be made between two theories, of which one is the more beautiful and the other obtains the better fit to the experimental data, he would choose the former over the latter.

It may well be true that the universe is a devious place. But we prefer the theoretical model of it which we construct in order to explain and predict its behavior to be a simple (and hence esthetically attractive) structure. This virtue of simplicity is the contribution of the human race to the laws of nature. To us it is most important, for it is precisely this property of a good theory which should enable us in a mental institution to distinguish the patients from the staff.

questions for the reader

There is considerable evidence associating diseases such as lung cancer with cigarette smoking. However, one can produce various alternative deductions from these data: (1) Cigarette smoking causes lung cancer. (2) Lung cancer causes cigarette smoking. (3) Both phenomena are independently induced by a third factor. (4) There is no real connection; the data are misleading or have been falsified. Develop these four arguments in light of the evidence, and then rate them according to their extent of *ad hoc* reasoning.

11   **The operational definition. When is a question not a question?**

Man is nothing else than his plan; he exists
only to the extent that he fulfills himself;
he is therefore nothing else
than the ensemble of his acts,
nothing else than his life.

JEAN-PAUL SARTRE
Existentialism and Human Emotions

Modern physics has been a process of restructuring our understanding of nature. Since we are a part of nature, it is only right that we should also reflect upon ourselves. This century has seen great changes in our way of thinking. In the affluent nations there has been an escalation of freedom to such an extent that people born more than twenty-five years ago are waiting in suspense for the other shoe (of repression) to fall. The arts have become dominated in large measure by a world outlook in which hardly a play or motion picture fails to convey what appear to be expressions of the futility and meaninglessness of life. The youthful cult of irreverence has destroyed all the sacred cows, and the educational institutions which were the fountainhead of the questioning process have themselves become a primary target. If the present age is anything it is an age of questioning.

We cannot blame all this entirely on the physicists. Yet it is somewhat paradoxical that the science of physics, itself a model of formalism and order, has for more than half a century been quietly injecting into the mainstream of human thought a philosophical outlook characterized by the weakening of traditional structure.

The theory of relativity did not take long to make its presence felt even among many people who had little to do with physics and undoubtedly had only the most general conception of what had been discovered. In 1913, only a few years after Einstein published his fundamental papers on relativity, a painting exhibited in a New York armory created an international uproar and established a new trend in art. It was called *Nude Descending a Staircase,* and the painter, Marcel Duchamp, has since been called the "grand Dada" and "spiritual father of the pop generation." According to his obituary in the *New York Times* of October 3, 1968, Duchamp was "especially appealing to more than a generation of intellectuals who felt that science had stripped them of traditional values and left them spiritually bankrupt." His work was described as "the expression of time and space through the abstract presentation of motion." It was characterized by apples which did not "condescend to the laws of gravity," the act of love as a "fourth dimensional ritual of machines," and an expressed disbelief in the meter as a fixed unit of measurement. "Why," Duchamp asked, "must we worship principles which in fifty or one hundred years will no longer apply?" One of his final works was made by drawing lines with templates he solemnly constructed by dropping pieces of thread precisely one meter long at random on a painted canvas from a height of precisely one meter. This last painting, called *The Bride Stripped Bare by Her Bachelors, Even,* was never finished. In this too perhaps he was a couple of generations ahead of his time; only ten years after the *Nude* was exhibited he refused to do any more work and stopped painting altogether.

There is no intended suggestion here that Duchamp's nihilism was universally shared by the painters who followed. And the relative merits of nonrepresentational as opposed to classical art forms are certainly not the issue. What is interesting is that art and science do not appear to exist entirely unto themselves.

The responses frequently evoked among non-scientists by Einstein's findings include such remarks as "Nothing is real," "Everything is relative," or "You can't believe anything, because it will probably turn out to be wrong anyhow." Such conclusions have in fact little to do with the actual state of things, and are only an expression of universal feelings of bewilderment and insecurity in the presence of what one does not know or fully understand. Yet some important changes had been introduced, and they were not restricted to the study of physics. While Duchamp was expressing his revolt in his art, the physicists in their turn were reviewing what had happened. It appeared that there was indeed something wrong with the way people had been reasoning.

However, one needs to distinguish between faulty logic and lack of information. Thus, for example, if a nation concludes that it has made some unfortunate policy decisions, what must it do to avoid similar mistakes in the future? There are several possibilities: (1) We may not have had access to all the correct information, (2) we may have proceeded on the basis of improper assumptions, or (3) we may have employed faulty reasoning. If it was information which was lacking, we take steps to improve our access to better information; but this probably does not represent a fundamental change. If the decisions were based on wrong assumptions, then obviously we must replace the old assumptions with better ones; although this may be more difficult, it requires only the courage to make the change. But if we were reasoning badly, then the correction becomes one of considerable substance.

The situation in physics after the turn of the century was that all three of these things had happened. Newton could not have been expected to know the results of the Michelson–Morley experiment, which came long after his time; there are undoubtedly more such surprises in store for us, and all we can do is continue the search. An incorrect assumption had been taken for granted about the addition of velocities according to the so-called Galilean transformation, namely, the belief that velocities in differential inertial frames had to be related by the rule of simple addition. All one could properly do about this was what Einstein proceeded to do—look for a simple alternative postulate which was in accord with the new evidence. But now it also appeared that our reasoning had been at fault.

To see how this happened, let us return again to the airplane stewardess. What does it mean to ask questions about her *absolute* velocity, i.e., with respect to nothing? We know very well what is meant by her velocity as it would be observed by the passengers in the aircraft frame of reference. We know what is meant by her velocity with respect to the balloon scientists, or the control tower operators, or the galactic observer somewhere in the solar system. We can say to these people, "Measure her velocity in your frame of reference," and they will all know what we are requesting, regardless of any practical difficulties they may encounter in carrying out the measurement. But how can we say, "Measure her velocity with respect to absolutely nothing at all"?

Yet this is effectively what Newton, one of the greatest scientists the human race has ever produced, was implying when he spoke of "absolute space" and "absolute time." We all nodded our heads, but neither he nor we actually knew what he was talking about. Unless space is something measured by some sort of meter stick, and time by some sort of clock, what are they? And a system of meter sticks and clocks is precisely what is meant by a frame of reference. Therefore space and time are meaningful only in terms of a frame of reference. A velocity likewise must be defined relative to something, since it can only be observed relative to something.

Thus one way to protect ourselves from making statements which have no meaning or which we do not ourselves understand is to define every concept in terms of an *operation*. The operation may be an experimental measurement, although the actual experiment may be difficult or impractical with existing equipment and be possible only conceptually in the form of a *Gedankenexperiment*. Furthermore, the operation should be uniquely identified with the concept, or there will be ambiguity in the definition. Similarly, a *question* becomes operationally meaningful if we can in principle define an operation (or experiment) which if carried out would give us the answer to the question. Thus Einstein carefully defined simultaneity of two events in terms of an operational measurement (by an observer placed midway between the events) before he permitted himself to ask any questions or reach any conclusions about this concept.

But suppose we cannot conceive how to measure something, and are therefore unable to offer an operational definition. Does it follow that what cannot be measured therefore cannot exist? Certainly not. Primitive man was probably unable to conceive the measurement of velocity, but to us velocity is very real. Indeed there must be a great many things which are inconceivable to us but which may be meaningful to posterity or to other intelligent beings.

What we are concerned with here is not the *correctness* of a concept, but only that what we think or say about it should have *meaning*. Whether we are talking to other people or to ourselves, we need to know what we mean when we make a statement or ask a question. The first step in finding an answer is to understand the question. And when we say, "What is the stewardess' velocity with respect to nothing?" we do not ourselves know what the question means. We might as well look very earnest and ask, "What is her predestinality?" We may even form a club consisting of people who incant this word periodically, throw their arms around each other, and insist that only they understand. But when asked to explain what they mean, they will say, "It is self-evident," or define it in terms of other words which are equally obscure. They may even say, "Your difficulty is that you don't try to understand. You must join our club and actually experience this thing in order to comprehend it."

Now there may be nothing wrong with belonging to such a club. Its members may have something important or pleasant or useful to themselves; they may even be ahead of their time. But whatever it is, it probably cannot be communicated in the usual manner, through the use of words. They are employing an idea which they are unable to define operationally. And the least one should ask is that they become aware that this is so.

The operational definition is not limited to concepts defined in terms of physical experiments. What is required is an operation, and it can even take a rather abstract form, as in mathematical definitions of concepts like continuity or analyticity. As long as the mental operation is readily understood, and can be repeated by large numbers of people, it meets the test.

Now it was a source of considerable surprise and embarrassment for physicists in the first decades of this century to discover that they had in effect unwittingly been members of just such a club as has been described here. And Newton, the greatest physicist of them all, had led the way. Neither he nor anyone else had ever properly defined absolute space or time or velocity; yet everyone had employed these terms, convinced that empty space must be given properties which would somehow support them. Then came Einstein's simple observation that the emperor has no clothes. It was like a bucket of cold water thrown over the human race, and physicists were not the only ones to get wet.

There is, however, an interesting complication. Although *constant* velocities cannot be observed without reference to something else, this appears not to be true of a *change* in velocity, which we call acceleration. As has already been pointed out, the passengers cannot determine the constant velocity of the airplane except with respect to something outside. But if the pilot suddenly puts on his brakes or fires a rocket engine, everyone will know it. The stewardess will end up in the lap of one of the passengers, and her tray and drinks will splatter against the bulkhead. There can be no mistake about it; the passengers do not need to have reference to something outside the window to know there has been a change in velocity, i.e., an acceleration. The same is true if the plane goes into a tailspin; everyone will feel it. Any acceleration of the aircraft frame of reference is readily detectable.

Thus one may be led to conclude that although absolute *velocity* has no meaning, absolute *acceleration,* or change in velocity, with respect to "empty space," is a very real thing. Does this mean then that our operational criterion has been bypassed, or that there actually exists an absolute frame of reference, at least for *changes* of velocity?

The operational requirement is not at all invalidated by this observation. In fact, it affords an excellent illustration of the principle. We can *define* an accelerated frame of reference by precisely the sort of experiences encountered by the passengers when the plane suddenly changes velocity or enters bumpy air. The panel lights which say, "Please fasten safety belts," are unmistakable testimony to the fact that the airplane is not expected to remain an inertial frame of reference. And the stewardess may have her operational definition, "Acceleration is when I find myself in someone's lap for no good reason."

However, the question of whether such operationally defined acceleration is *absolute,* that is, takes place with respect to empty space, or is *relative* to something, is more complicated. What we have glibly called "empty space" is not empty at all; it contains, among other things, the stars and galaxies. Hence, for all we know, it is acceleration with respect to the "fixed stars" which the passengers and crew are experiencing, and not *absolute* acceleration at all. Of course the passengers do not need to *see* the stars crossing their windows when the plane makes a turn or pitches over sharply in order to know that they are riding an accelerated frame of reference; they will know it from their spilled drinks or the tension in their safety belts. Thus what is operationally defined is acceleration with respect to "inertial space," and by this we mean a frame of reference tied to all the stars and galaxies in the universe. We say that it is the acceleration of the aircraft frame of reference *with respect to inertial space* which is associated with the sensations of the passengers. Such acceleration certainly has meaning in the operational sense.

But now we must face the question, what if there were no stars or galaxies? Would the airplane still be accelerating, and if so, would the passengers experience the same effects? This type of question was first posed by the physicist and philosopher Ernst Mach, who asked whether the presence of all the distant masses in the universe is the ultimate cause of the centrifugal force we experience on a merry-go-round, or the pressure we encounter against the backs of our seats when the vehicle we are riding spurts forward. In order to put the question in operational form, we now suppose that one of the governments of the world has finally attained its objective of Total Security, and has managed to develop the Last Great Weapon. This is a device which annihilates all the matter in the universe— stars, galaxies, everything. After a particularly successful test of the new weapon, the airplane passengers find themselves all alone in empty space. (Presumably the airplane has survived the cataclysm because of an effective air raid shelter program.) The pilot fires up his rocket engines, and the plane should therefore accelerate (whatever that means now). The question is, will the passengers still be thrown back against their seats when the rocket engines go on? We have no reason to expect that they will. Of course we do not know for certain, since no one has ever done the experiment. And we rather like to think that no one ever will. In the meantime all we can say is that operationally acceleration is defined with respect to inertial space, which presently *includes* the stars and galaxies.

The operational definition is of course not confined to physics. A great deal of time is often wasted in arguments about so-called controversial subjects—arguments which employ words meaning all things to all men, or perhaps nothing to anyone. Since most individuals rarely concern themselves with expressing their thoughts in a way which is operationally meaningful, they often find themselves reaching a state of anger or frustration because the other person "refuses to understand." One could easily find countless examples in the fields of politics, religion, sociology, etc. However, we choose instead a discussion of the sort which could take place on many a college campus; the subject is the use of drugs. The protagonists, whom we call Head and Square, are engaged in a controversy they are not likely to resolve quickly.

Head Your objection to drugs is based on lack of knowledge. Have you ever tried it?

Square I consider it unsafe.

H Everything is unsafe. Riding motorcycles, flying in airplanes, living in this century. There is only qualitative evidence; no one knows the probability of a bad trip, because no one knows how many people expose themselves. People say, "Look at those awful cases in mental hospitals. That's what drugs will do." But no one concludes because of the awful automobile accidents that driving cars should be illegalized.

S Even if I neglect the risk of a bad trip, or confine myself to something supposedly safe, like marijuana, I have no way of knowing how it will affect me, or what personality changes it will induce.

H And you won't know until you try it. How do you know you're not actually risking more by *not* taking it?

S Because I trust my mind when I'm straight. I could not trust my judgment if I were drugged.

H You may be right in distrusting your ability to drive a car when you are high. But I have observed an enhancement of certain faculties. There is unmistakable expansion of consciousness. Like the ability to appreciate music, or art, or poetry.

S I suspect that the primary effect is a loss of critical faculty. The ability to "appreciate" bad poetry is not what I would call expanded consciousness.

H   But I appreciate *good* poetry. And there is a general heightening of perceptiveness. I become aware of sights and sounds I never knew existed.

S   So does a person in a state of delirium.

H   But when a person is over his delirium he recognizes it as having been delirium. I gain insight which stays with me afterwards. Music which sounded particularly beautiful retains this special quality always. And the expansion is not limited to esthetics. I actually understand myself better as a result of these experiences. It's like instant psychiatry. I know people who were very confused and who straightened themselves out with only one or two trips on LSD.

S   I know some of the same people, and they don't look straightened out to me.

H   They're happy, which is what counts. They function. And they have resolved the issues which are relevant to them.

S   But how well do they function in coping with the external world?

H   And how well do you function in coping with the *internal* world? By turning on and turning off I can have it both ways.

S   What you have is a device to escape reality.

H   I don't take it to escape reality. I take it to get closer to reality.

S   You are introducing random disturbances in something like the human mind, which it took two million years to develop.

H   And one of the things developed in that two million years is the ability to take drugs. You are rejecting the experimental method. No one can assess an effect or understand it if he's never tried it.

S   According to you, only a person who is mentally ill is qualified to judge mental illness. I do not have to induce schizophrenia in myself in order to know its symptoms. I prefer to study the victim and retain my sanity.

H   You're assuming that the straight world is the sane one. If you were truly impartial, and someone showed you two societies— one embarked on a systematic form of mass suicide, and another which considers love and happiness to be the most relevant issues, which would you consider to be insane?

S   Are you suggesting I should take drugs to prevent war?

H   At least no one ever started a war while under the influence of drugs.

S   At the rate things are going, I wouldn't count on that record to last indefinitely.

The argument may go on interminably (and often does), but it is largely unproductive. Some of the key terms are: awareness, sanity, perceptiveness, reality, insight, unsafe, coping, consciousness expansion, relevant. But they do not necessarily mean the same things to everyone. In fact, it is more likely that they are used simply because they have pleasant or unpleasant connotations than that they have any definite meanings.

However, the discussion was not entirely vacuous. Head actually introduced the operational criterion when he challenged Square to define his use of the word "sanity." And Square did not let words like "perceptiveness" or "awareness" go entirely unquestioned. But he could have pressed more specifically for an operational definition, as follows:

Square What is "expanded consciousness"? Suppose just for the sake of argument that someone upon ingesting a particular chemical sees little green insects crawling on the wall—insects not seen by people who are free of this chemical. Can this person properly claim he has expanded his consciousness, that the insects are really there, but the drug has improved his ability to perceive them? If it is simply a matter of seeing things not seen by others, then he has certainly met the requirements. Or shall we instead consider only the *pleasant* sights induced by this chemical as evidence of an expanded mind, and reject the *unpleasant* ones? Do we call it "awareness" when we hear colors and see music, but write it off as a bad trip when the insects come on? Or shall we simply let the subject tell us himself when it is insight and when it is sickness?

Questions like "Does this drug expand my mind?" are simply not useful. They could more profitably be replaced by operational questions such as the following:

Does this drug increase or decrease the rate of heartbeat?

Does it stimulate the flow of adrenalin?

Does it induce sensations which the subject more often describes as pleasing, or displeasing?

Do people who use it tend to be "better" physicists (e.g., publish more papers, get more degrees, win more Nobel prizes, etc.)?

Do people who use it tend to be "better" musicians (e.g., play to larger audiences, sell more records, etc.)? One could of course ask whether such people "appreciate" music more, but this would be difficult to define, even in principle. One could claim, for example, that maximum appreciation is indicated by the person who most often repeats statements like, "Wow, I sure do dig that music," or who has the most beatific look on his face, or who simply listens to music most often. However, such criteria are by no means self-evident, and anyone wishing to discuss meaningfully the effect on music appreciation would be well advised to get such details straightened out.

Are people who take drugs more or less likely to raise families and hence procreate the race? (Note that no value judgment is implied; the overpopulation problem could raise some doubt about the "favorable" reply to this question.)

What proportion of adults who use drugs habitually hold down regular jobs, as compared to the population at large? (Again no value judgment is necessarily implied.)

Do drug users earn more or less money than nondrug users? (One may or may not regard the answer as interesting or significant.)

Do nations in which drugs have had the most extensive use tend to have higher or lower standards of living?

Do nations in which drugs are used extensively engage in wars more or less often?

Do students tend to get higher or lower grades after going on drugs?

What proportion of the people who use marijuana regularly go on to other drugs?

What proportion of people who have taken drugs are admitted to mental hospitals, as compared to a control group of the same general composition?

Answers to these questions may or may not be difficult to obtain. But at least we understand the question, and that is more than half the battle. "Does this drug expand my mind?" is somewhat like asking, "Does the earth move through an unobservable medium called the ether?" We cannot discuss either question meaningfully without operational definitions of the terms. In the case of mind expansion the situation is not altogether hopeless, because one could conceivably define an expanded mind as the ability to perform on some kind of examination, or to create something which evokes a favorable response in others, or simply as a pleasant sensation. Unfortunately one rarely specifies any of these things. Since there is no unique operational test, the question is ambiguous at best and meaningless at worst. What one usually "means" is that one "feels" expanded, and this is little more than a tautology. Of course in the case of the ether question the situation is worse, since the statement that it is unobservable *precludes* the possibility of an operational definition. It is interesting, incidentally, that Einstein never *denied* the existence of the ether. He merely said that there was nothing to be learned from contemplating the question.

Poet Your operational definition is an interesting idea, because it probably represents the epitome of the scientific method. I suspect scientists would love to turn us all into robots who assert emphatically that everything is hogwash when it can't be defined operationally. It just happens that we would then lose all ability to experience human emotion, or spiritual fulfilment, or religious conviction, or artistic expression. I suppose falling in love has no meaning either, unless you stop at every stage to ask the correct operational questions. As a matter of fact, I have a suspicion that even some of your scientific advances might never have taken place if you had insisted upon strapping everyone down in this way.

Scientist That is not the idea—to strap everyone down. There is no intention of excluding non-operational experiences. Everyone appreciates the differences between a computer and a human being. And I think you are probably correct in your suggestion that in scientific pursuits also one must be allowed a certain freedom to wander. As a matter of fact, even among scientists there is nothing mandatory about the operational definition. But one must somehow distinguish between science and pseudoscience, profundity and pontification, reason and wishful thinking. Otherwise, no matter how clever we consider ourselves to be, we are in danger of mouthing platitudes or chasing our tails. It is one thing to have an idea, another to examine it, and still another to communicate it. Once the dichotomy between reason and revelation has been made, everyone is free to make his own judgment. As for science, it could of course turn out that the operational criterion will prove inadequate for what lies ahead. But in that case we shall probably want something to replace it. In the meantime the very act of asking whether a term is operationally defined forces critical examination of issues and arguments which might otherwise turn out to be full of sound and fury, signifying nothing.

It could help to clear the air if politicians, administrators, and academicians were reminded of the usefulness of this concept, and asked on occasion to define their terms. However, as P.W. Bridgman pointed out, operational thinking is likely in the initial stages of its acceptance to prove to be an unsocial virtue. It is a rather powerful weapon, and must be used sparingly. Otherwise one is likely to find himself alienating his friends, who for their part may adopt a less than enthusiastic response to the suggestion that some of their supposedly profound observations and questions may actually be quite meaningless.

On the other hand, if one is to question anything at all he should be prepared to question his own statements. "The true meaning of a term," says Bridgman, "is to be found by observing what a man does with it, not what he says about it." Jean-Paul Sartre has his definition of a man, "You are nothing else than your life." The individual who claims undemonstrable insights and accomplishments says in effect, according to Sartre, "Circumstances have been against me. What I've been and done doesn't show my true worth. To be sure, I've had no great love, no great friendship, but that's because I haven't met a man or woman who was worthy. The books I've written haven't been very good because I haven't had the proper leisure. I haven't had children to devote myself to because I didn't find a man with whom I could have spent my life. So there remains within me, unused and quite viable, a host of propensities, inclinations, possibilities, that one wouldn't guess from the mere series of things I've done."

questions for the reader

Which of the following questions are meaningful in the operational sense? Supply definitions of ambiguous terms if this seems appropriate. Remember that what is at issue is not the answer but the question.

1. Is it possible that people are reborn after death, but without memory or evidence of their previous lives?

2. Can the color I call red look green to someone else, and vice versa?

3. Is there an end to space?

4. Can a physical theory be consistent with all existing evidence and still be wrong?

5. Is there such a thing as free will?

6. How can we prove we exist?

7. Is the statement, "Everything that goes up has to come down," a law of nature?

8. Is there a God?

9. Is it human nature to make wars?

10. Is it possible for a natural law to be broken?

11. Do drugs induce personality changes?

12. If I had my life to live over again, would I do things differently?

13. Suppose that an instant before you were killed in an accident, an exact "carbon copy" of you were constructed, with every genetic and biochemical detail exactly right, including all personality traits and memories of past experiences. Would you still be alive?

Now make up your own examples.

12   Waves and particles

I never saw a moor,
I never saw the sea;
Yet know I how the heather looks,
And what a wave must be.

EMILY DICKINSON
Time and Eternity, XVII

Dust had hardly settled over the unsuccessful search for a pre-ferred frame of reference when another major revolution started brewing. The new branch of physics which was the product of this revolution came to be known as quantum mechanics, and its chal-lenge to traditional thinking was in some ways more serious than that posed by the theory of relativity. It was the physicists them-selves who were most disturbed, since the experimental evidence was adding up to a model of the physical world which violated fun-damental assumptions about the very nature of knowledge and reality. To this day the new student of quantum mechanics often finds himself wavering between acute skepticism and the uncom-fortable feeling that he may be losing his mind.

In 1926 physicists gathered in Copenhagen for a conference to interpret their findings, and their conclusions were reviewed and reexamined at the Solvay conference in Brussels the following year. When they went back to their respective countries afterwards it must have been with some sense of discomfort and trepidation, since if they were to relate everything to their countrymen they were likely to be disbelieved at best; at worst in some cases political recrimination was a distinct possibility. Of the Copenhagen ses-sions Heisenberg says, "I remember discussions with Bohr which went through many hours till very late at night and ended almost in despair; and when at the end of the discussion I went alone for a walk in the neighboring park I repeated to myself again and again the question: Can nature possibly be as absurd as it seemed to us in these atomic experiments?"*

Today quantum mechanics has become the basis of much of modern engineering; it has revamped the science of chemistry, and has produced a new field of technology known as the solid state, which provides us with transistor radios and other electronic devices. But the wild philosophical concepts which were engendered by the discoveries of the first quarter of this century have reached the general public only indirectly. The "humanists" find the mathematics a bit too much, and the "practical" engineers prefer to avoid that universal heady feeling that one is losing his mind when he probes too deeply into quantum mechanical principles. But on some level of consciousness word has managed to leak out that the deterministic picture of the universe has broken down.

It is not clear what is the best way to begin. One could draw analogies to the Heisenberg Uncertainty Principle from everyday life; or one could make philosophical observations about one's inability to predict the future even with all available knowledge about the present. We shall, however, take the conservative approach, and begin with a description of some of the relevant physical phenomena.

Until the first quarter of the twentieth century it was believed that a particle is an entirely different thing from a wave. This statement may appear puzzling, and sounds rather like saying, "Until the twenty-first century people thought that a girl is not the same thing as the Pacific Ocean." It may be true, but why is it interesting?

Although hardly anyone is likely to confuse a girl with an ocean, things are not so simple when we consider a phenomenon such as light. Is a beam of light a train of waves, or is it a stream of particles (Fig. 12–1)? This question has worried physicists for hundreds of years. Newton examined the evidence then available and decided it was a stream of particles. If his theory was correct,

---

* From W. Heisenberg, *Physics and Philosophy,* New York, Harper and Row, 1958. Quoted by permission of the publisher.

then the light coming to us from stars millions of light years away consists of particles which have made the long trip through space. But after Newton's death the analysis of optical experiments presented conclusive evidence that light is a wave phenomenon.

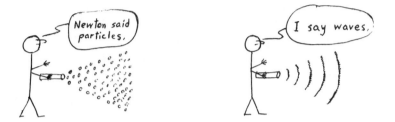

Fig. 12–1 The wave-particle question.

In order to understand what all this means, we had better first determine whether what we are asking is an operational question. We therefore list some of the differences (and similarities) between particles and waves.

A particle can generally be *localized* in space, whereas a wave is *spread out* over a region. This means that although one can generally say fairly precisely where a particle is located, in the case of a "pure" wave (mathematically called a sine wave) the best one can do is specify the location of the *peaks* of the wave, that is, the "crests" and the "troughs," the "tops" and the "bottoms." Thus our particle in the first picture of Fig. 12–2 is floating on one of the wave crests. The three scenes show this wave peak advancing to the right in time as it passes under the particle. The distance between two such consecutive peaks is called a *wavelength*. The height of the peak, on the other hand, is known as the *amplitude* of the wave. Just as a distinctive property of a particle is that its position is rather precisely limited in space, so wavelength and amplitude help to identify a wave, which is spread out in space. It is when the amplitude of offshore ocean waves becomes excessively large that the weather bureau issues small craft warnings.

Of course not everything in nature lends itself to such neat categorization. Sometimes a particle may be blown up into something big as a balloon, as in Fig. 12–3. Then it can no longer be so sharply localized or "limited" in space; we may even not wish to call it a particle any more, since its position is now defined only "on the average" or "for the most part" or "in general." Even when it is a small particle, there may be circumstances in which it trembles or darts about so rapidly in the vicinity of its "average" location that its position is blurred and becomes virtually impossible to determine precisely. In either case, since it becomes difficult to specify its position with great accuracy, we could say that our particle has acquired one of the aspects of a wave; i.e., it is spread out over a region instead of being concentrated at a point.

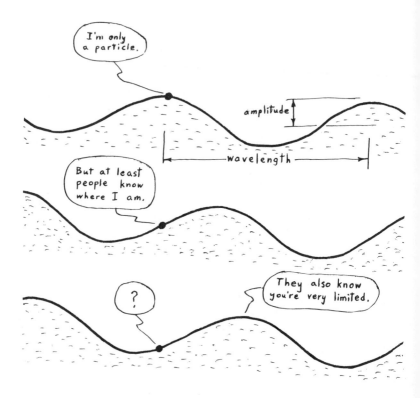

Fig. 12–2 *The particle and the wave.*

Fig. 12-3 *The balloon and the wave packet.*

By the same token, sometimes ocean waves pile on top of each other for a time (the mathematical term is *superposition*) in just such a way as to produce what the surfers call a "big wave" or the physicists call a *wave packet*. In the figure this temporary "packet" or "bundle" of waves is seen traveling to the right as it passes under the balloon. It is no longer the "pure" regular sine wave, but consists instead of a large peak preceded and followed by successively diminishing smaller peaks. We can synthesize such a wave packet by adding sine waves of different frequencies and amplitudes. Now neither the wavelength nor the amplitude of the resultant wave is precisely defined, since the peaks may not be evenly spaced, and their heights will vary. However, in exchange for the diminution of its wave properties, the wave packet has managed to acquire a degree of localization. In this respect it could be said to resemble a particle, since now at least people know where it is.

Poet If this is physics, then I am indeed prepared to believe you when you tell me not to call it an exact science. Are you planning to dissect every word in the language in this way, splinter all concepts, break down all distinctions?

Scientist That would be a rather unrewarding task. We only do this sort of thing when nature forces it upon us. But I am pleased to hear that you and I have found a basis for agreement.

Another property of a particle is its velocity. A wave has velocity also; but it does not mean quite the same thing. In the case of a particle everyone knows the operational definition of velocity is that you observe its position at two instants of time, and then divide the distance x between the two points by the time interval t. Thus the velocity V is distance (we sometimes call it *displacement*) divided by time,

$$V = \frac{x}{t}. \tag{12-1}$$

This is how the airplane passengers determined the stewardess' velocity.

In the case of a wave we would have to measure it differently. Let us return to the pure sine wave of Fig. 12–2, which has a precise amplitude and wavelength. We cannot determine the *position* of the wave, since this is undefined; but we can note the position of one of its peaks. Then we just stand there and watch them go by, clocking the time from one peak to the next. This time interval from one peak until the next corresponding peak is called the *period*. Dividing the wavelength λ (pronounced lambda) by the period T is again distance divided by time of travel, giving us a velocity, called the *phase velocity* of the wave. (The crest is called one "phase" of the wave, and the trough is a different "phase" of the same wave.) Thus phase velocity v is expressed by

$$v = \frac{\lambda}{T} ,$$ (12–2)

since λ is the distance traversed by the peak in the time period T.

The important thing to realize about a wave is that the *medium* (in this case the water) is really not moving forward. The particle floating on the surface of the water will bob up and down as the waves pass under it, but it remains in its original position. If the water itself were traveling forward, the floating particle would move forward along with it; but it does not. In the case of waves such as we have been discussing, which are called "traveling waves," what is traveling is not the medium but the message (if we may borrow a phrase from the opposition). The "message" in this case is the bump on the water which we have called the peak or the trough. In the case of a wave packet this bump is particularly easy to identify, since it is the biggest bump in the neighborhood. The presence or absence of the bump constitutes "information," and a series of them can convey a complex coded "message." In any case, although the bump travels forward, the water itself merely goes up and down.

Poet Then how does a surfer manage to ride the waves? According to what you say, he should simply bob up and down; yet the surfboard goes forward.

Even if he has not studied physics, the surfer learns to watch for the big breakers. This is why there are no surfers on the smooth waves of the deep ocean, but only in the shallows. Once a wave starts to break, local displacements and turbulences appear which propel the surfer forward. But strictly speaking the water then begins to lose its wave status, at least in the sense that we have defined it.

Of course what we see on the ocean represents only one type of wave. In the case of sound pressure waves, air molecules are driven toward the ones in front of them, causing pressure to build up locally and then to be relieved as the crowded molecules push against the ones still further up the line. It is the crowded condition of the molecules which constitutes the pressure; and what travels forward is this state of being crowded (not the molecules themselves), resulting in successive peaks of high and low pressure as the air contracts and expands. These pressure peaks are propagated forward, just like the peaks of ocean waves. In this case the pressure is the message and the air is the medium. Such a disturbance coming from the speaker of an audio system is depicted in Fig. 12–4a, and constitutes an air pressure (or compression) wave. The region in which molecules at a given instant are close together may be said to correspond to the crests of the ocean waves, and the wide spaces, where the pressure has temporarily relieved itself, to the troughs. (Of course real life molecules are not so neatly arranged in rows as in the figure, and move about in a random manner, but the net result is the same.)

One can even make up a "people" wave, consisting of a line of people (in Fig. 12–4b the people are the medium) who relay a message (in this case a tap on the head), which travels down the row with a characteristic phase velocity. The blow on the head may be regarded as the "crest" of the wave, and corresponds to the crowded state of the air molecules in a sound wave.

In both these examples it is the *state* of the medium which is being transmitted, not the molecules or the hammers; these are truly wave phenomena, not to be confused with Newton's particle model of Fig. 12–1, in which the "particles" of light were conceived as actually emanating from the flashlight and then moving away from it.

By the nineteenth century everyone was pretty well convinced that light consisted of waves, not particles. Overwhelming evidence to support this conclusion had been obtained, information which unfortunately had not been available in Newton's lifetime. We shall shortly examine some of this evidence, but first we ask ourselves the traditional question: If light is a wave, then what can be waving? The light comes to us after passing vast regions of rather empty space; what is the medium of propagation for light transmitted across a vacuum? What corresponds to the water of the ocean waves, the air of the sound waves, and the little men with hammers of the "people" waves?

Fig. 12–4 *Types of waves.* (a) Acoustic pressure wave from a speaker. (b) "People" wave.

In answering this question we shall avoid inventing a fictitious "ether"; we know by now what troubles this is likely to get us into. Instead we return to the "people" wave of Fig. 12–4b. Suppose that someone stands behind the last little man in the column and holds a very powerful magnet. If he is clever enough, he may, by waving this magnet up and down in just the right way, cause the hammers to move up and down without the active cooperation of the people in the column. It might take a great deal of practice, but it should certainly be possible to wiggle the hammers by waving the magnet. Furthermore, the hammers which are closest to the magnet will be the first to respond to this changing *magnetic field*, while the ones farther down the line will take slightly longer to get the "message." Thus a sort of "wave" disturbance ripples down the row of hammers, causing each one to rise and fall.

After practicing this feat for a while one probably grows tired of having to wave around such a heavy magnet, and may decide instead to replace it with a stationary electromagnet. The current in the electromagnet can be varied periodically by turning a small dial back and forth with the expenditure of relatively little physical effort. It is a bit tricky to manipulate hammers in this way, but given enough time and patience, one can learn to send a wave of hammer movements down the line of little men simply by twiddling the dial controlling the current in the electromagnet. The hammers are now said to be responding to a traveling *magnetic wave* propagating along the column.

Alternately, instead of using a magnet, we can deposit electric charges on all the hammers, and then wiggle another charged particle up and down at one end of the column, causing the electrically charged hammers to move up and down in response to the motion of the electrically charged particle. (If, for example, the hammers are charged positively and the particle negatively, there will be a force of attraction.) This time it will be an *electric* field moving the hammers, and a traveling *electric wave* propagating along the column of hammers.

Now it may be shown theoretically, and also demonstrated experimentally, that such a varying electric field always produces an accompanying magnetic field, and conversely a varying magnetic field always produces an electric field. This is due to the fact that electric and magnetic phenomena are simply different manifestations of the same thing; thus electric current in a wire wrapped around an iron bar constitutes a magnet, and rotating magnetic poles in a generator produce electric power.

Such waves as we have just been creating are therefore called *electromagnetic waves*. Electric charge is caused to move up and down an antenna wire in a radio station, transmitting precisely such waves to distant radio and television receivers. The frequency f (in *cycles per second*) with which a moving electric charge passes up and down an antenna wire is also the frequency of the transmitted wave, corresponding to the number of times per second that our hammers rise and fall; and the period T is the number of *seconds per cycle,* or the time it takes to complete one rise and fall. Thus

$$f = \frac{1}{T} \qquad (12\text{--}3)$$

is the relation between the frequency and the period of the wave.

Of course actual radio frequencies are much too high (millions of cycles per second) to move such heavy hammers, but the principle is there. In the case of visible light, which is exactly the same type of electromagnetic wave, the frequencies to which our eye responds are even higher, being measured in 10^{15} cycles per second (a one followed by 15 zeros).

The phase velocity of electromagnetic waves is given by the same equation (12–2) which we had for ocean waves. We can further combine this with equation (12–3) above, relating period and frequency. The phase velocity of light or any electromagnetic wave (let us call it c to distinguish it from the v of the ocean wave) is then

$$c = \frac{\lambda}{T} = \lambda f. \qquad (12\text{--}4)$$

Thus the phase velocity of light is equal to the wavelength times the frequency. This phase velocity is the same universal speed of light *in vacuo* which was predicted by Maxwell's equations, and which we encountered in our discussion of relativity and the Michelson–Morley experiment. It amounts to 3×10^8 meters per second, or 186,000 miles per second. Thus when we talk to our astronauts on the moon, some 230,000 miles away, it takes light or radio signals well over a second to make the trip, and it takes more than $2\frac{1}{2}$ seconds to get the reply to a question. Such conversations are likely to be good training for people who are in the habit of interrupting rather than listening. They will also help to develop patience in preparation for future conversations with, say, a planetary system of the Great Nebula in Andromeda, which will require a wait of some four million years before receiving the answer to a question.

Poet It appears to me that you have done exactly what you objected to in the case of your ether theory friend. You have invented a medium, only you don't call it the ether; now it is little men with hammers.

Actually, it is about time we got rid of them too. They have out-lived their usefulness. Their movements indicate the *presence* of the changing field, but they are themselves not necessary in order for the field to exist. If we fire all the men except the one at the end of the line, who is farthest away from the spot where we are waving around our magnets and electric charges, the motion of the one remaining hammer will still continue. It works even if we evacuate all the air between the transmitter and the receiver. The hammer still moves. Physicists in days gone by gave this effect the name "action at a distance." Nowadays we usually describe it as a field of force, with electromagnetic waves which travel through empty space between the "source" and the "receiver." It thus becomes ap-parent that such a field does not require a physical medium. The only function of the men and hammers was to enable us to *picture* the electromagnetic field. Now that we know it is there, we can even get rid of the last remaining man and hammer. In our newly found ability to abstract we have attained a state akin to Emily Dickinson's; for we "know how the heather looks, and what a wave must be."

13   Light is a wave

Put out the light, and then put out the light:
If I quench thee, thou flaming minister,
I can again thy former light restore
Should I repent me; but once put out thy light,
Thou cunning'st pattern of excelling nature,
I know not where is that Promethean heat
That can thy light relume.

WILLIAM SHAKESPEARE
Othello, Act V, Sc. 2.

In 1801 there was discovered unmistakable evidence of the wave nature of light. Until that year Newton's corpuscular (particle) theory had prevailed. The man who established the wave theory beyond a reasonable doubt was Thomas Young, and his demonstration was so simple that it may fairly easily be repeated by the reader. In our discussion of particles and waves we have as yet made no mention of the most interesting difference between them—their respective rules of addition.

If you send a small boy to the grocery store with instructions to buy a quart of milk, and then do the same with another small boy, the chances are fairly good that you will get back two small boys and two containers of milk. The reason for this is that small boys and milk containers are rather in the nature of particles, and therefore add numerically. If they were more in the nature of waves, you might instead wind up with a large boy and a two quart container of milk, or alternatively with nothing at all. This is illustrated in Fig. 13-1. Two waves of the same amplitude and wavelength, when superimposed (added) *in phase,* i.e., crest to crest or trough to trough, result in a wave of twice the amplitude. This is called *reinforcement,* or *constructive interference.* However, when they are

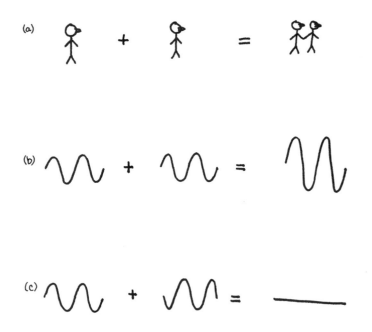

Fig. 13–1 *Addition of particles and waves. (a) Particles. (b) Waves (in phase). (c) Waves (out of phase).*

added *out of phase,* i.e., crest to trough and trough to crest, they simply cancel each other. This is known as *destructive interference.* Thus waves are subject to the principle of *superposition.*

   On the other hand, it is the nature of particles and people that they are not subject to superposition. For one thing there are no negative persons, or we would not be having a population explosion. Of course objects classified as particles may cease to exist in the sense that they can be made to lose their original identity; thus a pencil mark may be erased, or a human being may be destroyed. This tends to be a rather permanent state, however, and bears little resemblance to the process of cancellation which is obtained in the superposition of waves. Thus although people and particles can be destroyed, they cannot be cancelled; this is a direct result of the fact that they add numerically, not algebraically. The property of interference is exclusively a wave phenomenon.

Some rather interesting and surprising effects may be observed in the interference of waves. One of these is illustrated in Fig. 13–2. People holding opposite ends of a rope (or long spring) snap their respective ends so as to send identical but oppositely phased symmetrical rope waves (we should rather call them pulses or wave packets) traveling in opposite directions. Ideally the rope should be elastic, and is held tightly enough to take up any slack. The two pulses move toward each other, preserving their original shapes all the while, until finally they pass through each other at the middle of the rope and continue on beyond. At the instant they are superimposed the waves experience destructive interference, and both disturbances have apparently completely disappeared. But a moment later they reappear like a phoenix rising from the ashes and resume their travels down the rope. The interesting thing is that during the "moment of silence" the rope has to remember not only the fact that there were two pulses traveling in opposite directions, but also all the information about their original shapes, so as to be able to recreate them precisely as they were before the cancellation.

If you wish to try this demonstration, you may find it easier to lay the rope on a long table, and have your assistant and yourself snap the two ends right and left, respectively, rather than up and down. This will eliminate the asymmetry introduced by the earth's gravitational field as a result of the weight of the rope.

The important thing about the experiment is that it will not work for particles. Two small boys riding their bicycles head on cannot be expected to pass through each other successfully, and the result is rather more likely to be catastrophic. Of course this could be done in principle if all the atoms and molecules of which one of the boys is comprised just happened to line up in such a way as to pass through the intermolecular spaces of the other. But we should have to wait a long time and use up a large number of small boys before managing it. Even then at the moment of passing through we actually expect to see not cancellation, but only a very dense boy.

It may have been considerations of some such nature which led Young to try his interference experiment, in order to settle once and

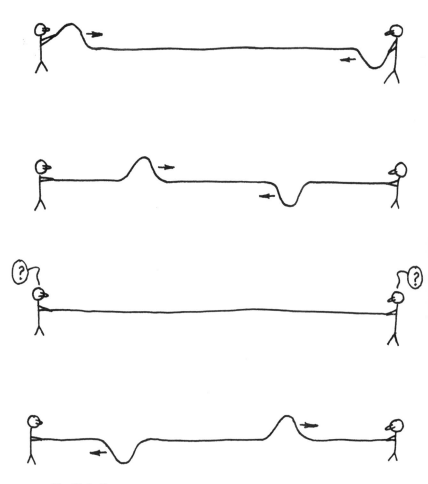

Fig. 13–2 Rope waves.

for all the question of whether light is a wave or a particle. It may be done not only with light, but with water waves as well. The effect is essentially the same.

We begin with a tank of water, in which circular waves are induced by poking two rods (or fingers) into the water at periodic intervals. Such movement excites circular outgoing waves from each of the points of insertion. The up-and-down movements of the two rods must be synchronized and have a fixed frequency, or repetition rate. The waves induced at these two source points are then called *coherent*, which means that both the frequency and the relative phase of the sources are fixed (not subject to random or unpredictable change), and the two sources are thus always locked together in phase. The distance between sources should be larger than the wavelength.

The interference pattern obtained from two such coherent sources is illustrated in Fig. 13–3. The arrows point to the spots where the rods pierce the water, and the two systems of concentric circles are the resulting outgoing circular waves. Solid circles represent at a particular instant the crests of the waves, and the dotted circles the troughs. Wherever a crest from one source crosses a crest from the other, or a trough crosses a trough, such an intersection is marked by a black dot. These peaks are points of *constructive interference*, corresponding to the superposition in phase of Fig. 13–1b. But when a peak crosses a trough, this is *destructive interference*, or cancellation of the sort represented in Fig. 13–1c; these points are marked by white dots. It may be seen that as the waves travel outward, so do the black and white dots. An observer lined up with the black dots will see large alternating crests and troughs (twice the amplitude of those coming from a single source), while an observer looking down the line of white dots will see a calm surface. The white dots constitute what is called a line of *nodes*; a node is a point on a wave midway between the crest and the trough, and retains the same water level it had before the wave came along. Of course if the two sources are precisely out of phase instead of in phase, the peaks will become nodes and the nodes peaks; this will, however, still constitute an interference pattern. What is important is merely that the relative phase remain fixed; if it changes back and forth, the pattern will be washed out. Incidentally, unless you do this experiment on a large lake, you may have to contend with reflections coming from the edges of the tank, which can complicate the picture once such reflections have made their appearance. "Official" ripple tanks are supplied with baffles at the edges to prevent reflection.

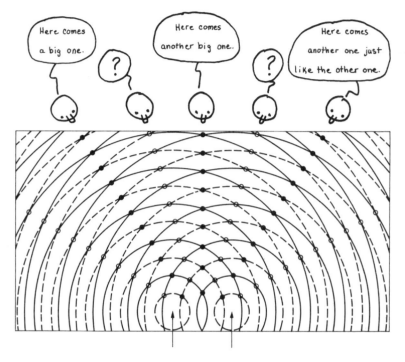

Fig. 13–3 *Interference of waves.*

The interesting thing is that if we perform the corresponding experiment with two coherent light sources, we get precisely the same result. This is seen in Fig. 13–4, which shows the interference pattern produced by coherent light passing through two slits which then proceed to act as wave sources, just like the rods which are inserted periodically in the ripple tank. In the case of light interference this phenomenon is usually called *diffraction*; when conditions are idealized by arranging to have the original light source and the screen (or photographic plate) effectively far from the slits, it is called *Fraunhofer diffraction*. The bright lines (or "fringes") in the figure are called *maxima*, and correspond to the black dots we used to label the points of reinforcement in the case of the water waves. The greater the amplitude of the resulting peaks (crests or troughs) of the light waves reaching the screen, the brighter are these maxima. (The intensity of the light actually turns out to be proportional to the *square* of the amplitude.) The dark lines correspond to the location of the nodes, where there is destructive interference, or cancellation; at these *minima* it is as if someone has "put out the light," but as a result it shines ever so much more brightly at the positions of the maxima.

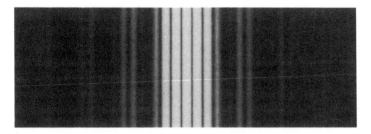

Fig. 13–4 *Fraunhofer diffraction pattern for a double slit. (From Bruno Rossi,* Optics, *Addison-Wesley, 1957.)*

In Fig. 13–4 monochromatic light was used, rather than white light coming from the sun, which is a superposition of all the colors. Each color in the solar spectrum corresponds to a different frequency (and wavelength); hence the diffraction pattern of white light produces a rainbow effect, due to the fact that maxima and minima for each frequency fall in different places on the screen. But with monochromatic light the delineation is quite sharp. The space between slits, the distance to the screen, and the wavelength of the light uniquely determine the interference pattern. A simple geometric calculation (see Appendix) obtains the distance in the diffraction pattern between the *central maximum* and the *first minimum;* this distance turns out to be

$$D_1 = \frac{\lambda f}{2d}, \tag{13–1}$$

where λ is the wavelength, d the distance between the slits, and f the distance to the screen.

This may be a good time to remind the reader of the relationship between frequency and wavelength as introduced in equation

(12–4). We now rewrite it

$$\lambda = \frac{c}{\nu} \, , \tag{13–2}$$

having taken this occasion to switch from the symbol f, used for frequency of a wave in the last chapter, to the Greek letter ν (pronounced nu), which designates the frequency of a light wave, in accordance with prevailing custom. Thus the greater the frequency of the light the shorter the wavelength (since c, the speed of light, is constant). The visible spectrum ranges from violet, with a wavelength of 4500 Angstroms (abbreviated Å), to red, with a wavelength of 6100 Å ($1Å = 10^{-8}$ cm; $10^{-8} = 1/10^8 = 1/100,000,000$).

There is, however, a problem in reproducing the ripple tank interference effect with two light sources. In the ripple tank we were able to control the way the rods went up and down and hence to keep them moving together. But light waves are produced by processes which take place in individual atoms or atomic nuclei. A light source consists of a large number of elementary *oscillators*, or charged particles moving rapidly to and fro. They send out random bursts of electromagnetic waves, and there is no means of controlling their phase. When there is no coherence (consistent phase relation) between the two sources there can be no systematic interference and hence no diffraction pattern. It is possible that the first time Young tried the experiment he simply passed light from the sun through two closely spaced pinholes (Fig. 13–5a), which should then have become sources of outgoing waves just like the rods piercing the water; but if this is what he did the experiment was a failure and produced no interference pattern. He might have given up at this point, concluding that Newton had been right, and that light consists of particles, not waves. But since he was aware of the fixed phase requirement for the two sources, he devised a very simple but clever means of solving this problem. It is, however, rather easier to do it than to understand why it works. Before passing the light through the pair of pinholes he first passed it through a single pinhole, as in Fig. 13–5b. The light emerging from the first hole spreads in spherical wavelets on to the second barrier, which should be a long distance from the first barrier compared to the space between the two holes. (The fact that waves spread in this way upon passing through

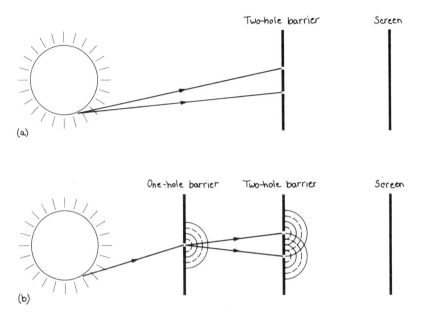

Fig. 13–5 *Young's experiment. (a) Incoherent light from the sun, consisting of many oscillators, with different path lengths through the two holes for any one oscillator. (b) Spatially coherent light obtained when the sun is effectively reduced to a point source. Now the path lengths to the two holes are essentially equal.*

a narrow passageway may be easily verified with water waves in a ripple tank.) The first barrier insures that light coming from any single oscillating source of electromagnetic waves in the sun travels essentially the same distance to either hole of the two-hole barrier (as can be seen in the figure); hence there will always be a fixed phase relation upon arrival at the two holes. Such light is said to be *spatially coherent*. The trouble with the arrangement of Fig. 13–5a is that light waves from an oscillator in one edge of the sun have *different* path lengths to the two holes and hence generally arrive out of phase, whereas when they come from an oscillator in the center of the sun the paths (not shown) have the same length. Such light is said to be *incoherent,* since it produces different interference patterns (corresponding to different phase relations at the two holes) for all the various oscillators in the sun, with bright lines from one oscillator appearing at the same portion of the screen as dark lines from the other. In such a case the maxima and minima will all be averaged out, and there will be no interference pattern. But in Fig. 13–5b both slits of the two-hole barrier receive equal treatment, and no matter in which part of the sun the light originated, the path lengths to the two holes are effectively equal (or at least miss being equal by a fixed amount). Thus even though the countless oscillators in the sun are randomly phased with respect to each other, the phase relation of the light waves in the *two holes* is the same for every single oscillator, and all such oscillators therefore produce the same interference pattern on the screen. This problem of coherence is not understood by everyone, and if it confuses the reader he may be assured he is in good company.

The preliminary one-hole barrier is a minor detail in implementation, but if Young had not been clever enough to introduce it, a great deal more time might have elapsed before the human race discovered that light is a wave phenomenon. Nowadays we get better results with slits instead of pinholes, and a lamp with a long straight filament acts as both source and first barrier. A piece of red or green glass makes the light monochromatic. The double slit barrier is made by scratching very thin straight lines on an opaque photographic plate or carbon-blackened piece of glass. The slits should be separated by something less than a millimeter because of the short wavelength of visible light. You hold the barrier in front of your eye, and look at the lighted filament through the double slit; what you see should convince you beyond any doubt that light is a wave.

Poet I have tried to be patient, but I hope you are not expecting me to listen to these technical details indefinitely. It has been mildly interesting, but after all I am not an optician or a radio repairman or a lifeguard, or whoever it is that is supposed to be interested in waves. I do not regale you with poetic techniques or stories about the influence of Chaucer's Canterbury Tales on the Elizabethan drama. So why are you doing this to me?

It would have been possible to replace this entire chapter with the statement that there is incontrovertible proof of the wave nature of light. But we are moving toward the climax of our story. It is important to realize the care and precision which were required to establish that light is a wave. If I were to skip the details, and instead introduce quantum mechanics by saying we have always assumed that there is in all things an "objective reality," with material objects which exist and move independently of any observer, and that now this assumption has apparently been violated, the reader would probably either misunderstand or refuse to believe it. If someone were to say, "The world as we knew it has just come to an end," you would probably say, "That's an interesting thought"; but really you would write it off as a figure of speech, or simply a dramatic way of calling attention to something far less catastrophic that you have probably known for some time, and have simply not bothered to verbalize. But if he said, "Come and have a look; I am not kidding; here is the proof," this would be an altogether different matter.

Soon we shall be standing where the physicists were in Copenhagen in 1926. The failure of classical materialist philosophy is a bitter pill for most people to swallow, and it is important to know its experimental basis. Otherwise the physicists will actually be sitting on this information alone, with no way of communicating it to the rest of the world, which must then feed only on rumors and half truths.

Poet It may be a bitter pill for the scientists, but we never believed in it in the first place. You imported this materialism along with your Industrial Revolution, but poets never rejoiced in it. We recognized the role of scientists in collecting and disseminating "objective" facts and building a technology, but still the poet or artist continued to search for truth in his own way. Without the poetic imagination and subjective vision of the whole there would be only charts and statistics and assorted data about pulse rates and blood counts, but there would be no human beings. Perhaps what has happened is that in your pedestrian way you are beginning to come around to conclusions which the rest of us have held all along. The physical approach to philosophy is evidently the laborious one. While you are cutting a path through the wilderness with your machete and examining every leaf of every branch of every tree with a magnifying glass, the poet is flying overhead and surveying the forest. When you have finished your detailed examination you say, "Aha, it's a forest." But to the poet it has been apparent all along.

Scientist If I may extend your analogy, it is not at all clear that you would be able to fly over the forest unless someone had previously constructed an airplane by the same methodical procedure you are inclined to deprecate. In any case you are no more able to live in the world or search for truth without the scientists than they can remain indefinitely uninfluenced by what you do. You may think you knew it all along, but I do not believe it. The human race has generated many philosophical concepts, as well as an infinite variety of superstitions, all of which were based on "vision of the whole." But only those ideas which have met the test of rigorous scientific examination have managed to survive.

P   If I understand you correctly, one of the "rigorous" ones apparently failed to make it also.

questions for the reader

1.   Does the disappearance of the wave pulses in the third scene of Fig. 13–2 violate energy conservation?

2.   How does the rope "remember" the shapes of the pulses when they have momentarily disappeared?

3.   Name some concepts which have survived the test of time despite their purely intuitive or instinctual origins, and make a prognosis of their chances for survival in the foreseeable future.

4.   It has been said that except for technology the human race has made little or no real progress since primitive times. The ancient Greeks believed that a stone dropped from the stern of a moving ship falls straight down into the water. What fundamental change in the human approach to problems does this suggest has taken place since those days?

14   A wave is a particle

Two voices are there: one is of the sea,
One of the mountains; each a mighty voice.

WILLIAM WORDSWORTH
Thought of a Briton on the
Subjugation of Switzerland

Just one hundred years after Young carried out his famous experiment it was discovered that Newton had been right all along—light consists of particles. In view of the previous insurmountable evidence that it had to be waves, this was understandably a rather awkward thought for people to contemplate. However, just as had been true in the case of the wave theory, once the facts were known there was no denying the conclusion.

In order to appreciate the importance of this surprising turn of events it will again be necessary to understand its experimental basis. One must not underestimate the significance of the paradox. The waves which produce interference patterns are spread out over virtually the entire apparatus, as may be seen in the water wave model of Fig. 13–3. It is in fact precisely this *spreading out* which makes the interference phenomenon possible. On the other hand what we mean by a particle is exactly the opposite; namely, it is *not* spread out, but is instead highly localized. It is indeed a contradiction to claim that something is everywhere, and then also to claim that it is concentrated at a single point and nowhere else to be found. Thus nature was really speaking to us in two voices; one said waves and the other said particles.

Despite the fact that a photographic plate seems to hear the wave voice, it turns out that the light meter mounted on top of the camera responds instead to the particle voice; the relevant phenomenon is called the *photoelectric effect,* and the man who first explained it in 1905 was Einstein. (It was the same year in which he presented his special theory of relativity; yet it was not relativity which won him the Nobel prize, but the photoelectric effect.)

This phenomenon was first observed by Hertz in 1887, and consists of the fact that ultraviolet light shining on a metal has enough energy to knock atomic electrons right out of the surface. That the ejected particles are actually electrons was established by Lenard in 1900, when he measured their ratio of charge to mass and found it to agree with that observed for electrons in other experiments. But in the course of his measurement some rather perplexing details were noted. The general idea is shown schematically in Fig. 14–1a. A glass tube from which air has been evacuated contains two copper plates, or electrodes, which are connected to a battery through an ammeter (which measures electric current). The electrode on the left (connected to the negative terminal of the battery) has been polished to a smooth clean surface; light of some frequency in the ultraviolet range is permitted to fall upon the inner surface of this electrode, which is called the *photocathode.* (The outside of the tube is opaque to light except for a region which is left transparent for this purpose.)

The battery is varied from several (positive) volts down to zero, and then the battery terminals are reversed, applying what we would call negative voltage, i.e., opposite to that represented in the figure. The resulting current measured by the ammeter is plotted in Fig. 14–1b. Such current is simply the movement of electrons down the length of the wire from the negative terminal of the battery to the positive terminal. This flow normally cannot take place in Fig. 14–1a because the circuit is interrupted by the space between the two electrodes. However, when light falls on the photocathode (the "negative" electrode), the negatively charged electrons are ejected from its surface and attracted to the other (positive) electrode, whereupon the circuit is closed and current begins to flow. When the light is turned off the current ceases. When it is turned on again the flow immediately resumes. Thus clearly the light is making possible the flow of current.

The amount of current is, however, independent of the applied positive voltage of the battery, as shown by the horizontal portion of the curve in Fig. 14–1b; it increases only as the *intensity* of the ultraviolet light is increased. (This is not shown in the figure, but it would simply raise the horizontal part of the curve.) When the battery voltage is reversed (plus becoming minus and minus plus), current still continues to flow, decreasing, however, until the voltage reaches a maximum negative value (a couple of volts or so), designated as V_{max} in the figure, whereupon all flow ceases. Beyond this negative voltage there is no current *regardless of how intense we make the light.*

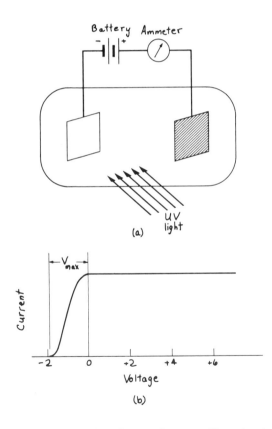

Fig. 14–1 *The photoelectric effect.* (a) *Photoelectric cell.* (b) *Current vs. voltage.*

There can be no question but that the light is knocking electrons out of atoms in the photocathode, and that as long as there is
any positive voltage at all, these electrons are being swept up by the
electric field created by the battery and driven toward the positive
electrode.   Increasing the intensity of the light leads it to knock
out more electrons, and therefore causes more current to flow.
When the applied voltage is reversed, it tends to drive the electrons
back into the (now positively charged) photocathode, and only the
most energetic ones manage to reach the other electrode; even these
cannot make it once the back voltage reaches V_{max}.

But there are two big surprises in this experiment.

Surprise 1

If you try to knock electrons out of a copper electrode with red or
orange or yellow or green or blue light, or with any wavelength
longer than that of ultraviolet, there is simply no photoelectric effect.
It does not matter how high you make the assisting positive voltage,
or how intense the light.   The weakest source of ultraviolet light
will do it nicely, but nothing happens with blue or red light, even if
it is brighter than a thousand suns!   However, if you now switch
to a *potassium*-coated photocathode, then it works for either blue
or ultraviolet light, but will still not work for red, orange, yellow,
or green.   Each metal appears to have its *threshold* wavelength, and
refuses to respond to longer wavelengths.   Furthermore, the negative voltage $-V_{max}$, which cuts off all the current and is therefore
proportional to the energy of the most energetic electrons, is the
same for one particular metal and color of light regardless of how
intense we make this light.   Thus the existence of the photoelectric
effect and the maximum energy of the ejected electrons depend only
upon the color (frequency or wavelength) of the light, and not at
all upon its intensity.   This is like saying that a ship cannot be broken
away from its mooring no matter how big the tidal wave which en-
gulfs it, whereas a little boy making waves by poking his finger in
the water at a high enough frequency can send it right up into the sky!

Surprise 2

When we switch on light of an appropriate color, the photocell
responds almost instantly.   Now we happen to know the approxi-

mate radius of an atom (it is of the order of one Angstrom, or 10^{-8} centimeter) and hence its cross-sectional area; thus we can calculate the rate at which light of a given power (energy per unit time) falls upon the atom when it is placed in the path of such light. We can also calculate the energy of the most energetic electrons by knowing the cutoff voltage $-V_{max}$ it takes to stop them. When we do this for a typical one-watt source of light, for example, it turns out that it should take something like a minute of exposure before any atom in the photocathode can absorb enough light to send out an electron of the observed energy. Yet the actual response time of such a photoelectric cell to the turning on of the light is less than 10^{-9} seconds (one-billionth of a second); in other words it is practically instantaneous. How does the electron find energy enough to be ejected long before the atom has had time to absorb this quantity of energy from the light source? Why is the little boy able to flip the big ocean liner out of the water with the very first touch of his finger?

Einstein decided that in order to explain the photoelectric effect he would have to give up the idea that light is a wave. This was a rather wild thing to suggest, but he was a young man, and wild ideas appear to be the special province of the young. Unlike the patriarchs of old, it is mostly the young to whom God in the twentieth century evidently prefers to address Himself—at least such would seem to be the case in the world of physics.

Einstein took his cue from the work of Planck, who had analyzed the spectrum of light radiated from bodies as a result of their temperature; physicists call it *blackbody radiation*. (A "blackbody" radiator happens to lend itself to such study because it never reflects light, but only absorbs and reemits.)

Everyone had been trying without much success to find a reasonable mathematical model which could account for the particular distribution of colors seen in such radiation. Planck discovered that if he assumed the walls of the radiator to contain oscillators with precisely distinct (scientists call it *discrete*) values of energy, rather than the continuous range of all possible energies, he could compute exactly the radiated spectrum which was being observed. Now the energy of an oscillator depends on how far the particle swings in one direction before it has to reverse itself and swing back in the other, much like a child on a swing or the bob of a pendulum;

the wider the swing the higher the energy. For every frequency in the electromagnetic spectrum, i.e., for every frequency of "swing," Planck assumed that the oscillators were allowed to have only those energies which were whole number multiples of a particular *quantum*. It was as bold a conjecture as proposing, for example, that on some strange planet all the people have heights measured in whole inches, but never in fractions of an inch; presumably a child on such a planet remains at the same height instead of growing continuously, until on an appropriate day he "pops" an entire inch, after which he retains this new height until it is again time to "pop."

Planck postulated that for some mysterious reason oscillating energies came only in integral (whole number) multiples of the quantity $h\nu$, where ν is the frequency of the oscillation in cycles per second, and h is a universal constant of nature, now known as Planck's constant. In other words, the elementary oscillators which radiate electromagnetic waves of a given frequency were conceived to be charged particles moving back and forth a particular number of times per second (the frequency ν), but having only energies

$0, h\nu, 2\ h\nu, 3\ h\nu, 4\ h\nu, \ldots ,$

and so on. This state of affairs is illustrated in Fig. 14–2. None of the oscillators is allowed to have energies like $3\frac{1}{2}\ h\nu$, or $7.6\ h\nu$, any more than the people on the Peculiar Planet are permitted to have heights measured in fractions of an inch.

There is certainly no plausible reason for such a conjecture; why nature should want to be so contrary was probably as much of a riddle to Planck as it is to the reader. It is just that he happened to notice when he introduced this assumption that he obtained an equation for total energy radiated by a body at all its various frequencies which was in precise accord with what had been observed experimentally; whereas with the entire continuous range of energies no relationship could be found which would work. He was thereby able to compute the magnitude of Planck's constant h by fitting it to the data on blackbody radiation. It turned out to be $h = 6.6 \times 10^{-34}$ joule-seconds, a very small number, since it has in the denominator a one followed by 34 zeros.

Einstein seized upon this strange conjecture to make an even stranger one. He said that regardless of what Planck's oscillators themselves were doing, one thing seemed evident; the electromagnetic energy (light) which was being emitted by a hot body was coming out in integral multiples of $h\nu$. As an oscillator loses energy, it apparently drops discrete *quanta*, or chunks of energy of magnitude $h\nu$, just as the people on the Peculiar Planet grow in bursts of height of precisely one inch. After losing such a quantum the oscillator evidently pauses for a while before being able to emit another.

Fig. 14–2 *The property of discreteness (a) The peculiar planet. (b) Planck's oscillators.*

Such discrete bursts of energy are what appear in the radiation spectrum of a blackbody; and they serve to explain the photoelectric effect. The light falling on the photocathode consists of this very same kind of electromagnetic radiation, since the sun is a hot radiating body. Thus an electron in the metal, when illuminated by light of frequency ν, absorbs one of these energy bundles and is ejected from its atomic dwelling place.

It is somewhat like knocking bottles off a fence by throwing rocks at them. A bottle falls only when it is hit, and it leaves the fence with essentially the energy it was able to absorb from this one rock. Increasing the intensity of the light corresponds to throwing more rocks in a given time; more bottles leave the fence, but each bottle still acquires the energy of only one rock. Increasing the frequency of the light corresponds to using bigger rocks, and bigger rocks permit you to knock off bigger bottles. But throwing rocks *more often* (increasing the intensity of the light) will not help you get any bottles if the rocks are too small; it only serves to get *more* bottles if the rocks were large enough to do any good in the first place.

Einstein said that light quanta of energy $h\nu$ reaching the surface of the photocathode were knocking electrons out of their atoms. In order to get the electron out so that it can be swept up by the electric field produced by the battery, this energy has to be large enough to overcome the attractive force of the atoms comprising the metal, just as a space vehicle has to receive a large enough burst of rocket power to escape the earth's gravitational field. In the case of a copper photocathode only frequencies ν in the *ultraviolet* region provide big enough bursts of energy $h\nu$ to do the job, whereas with certain other metals light in the visible region is sufficient. Of course, some of the electrons are knocked out by light quanta which have managed to penetrate *below* the surface of the metal, and such subsurface electrons lose some of their energy just in the process of reaching the surface; these may still escape if there was enough energy $h\nu$ provided them in the first place. The negative cutoff voltage $-V_{max}$ which drives *all* the electrons back into the metal, finally cutting off the photoelectric current, is a measure of the energy of the most energetic electrons, those knocked out at the very surface of the metal. This explains why increasing the intensity of the light (throwing more rocks) effects no change in this maximum energy of the ejected electrons, and hence in the magnitude of voltage V_{max} which can stop them.

Furthermore, we are now in a position to explain Surprise 2. The electron in the metal is like a block floating in a bucket partially filled with water. Only when the bucket overflows can the block "escape." If light consisted of continuous waves, like the stream of fluid emerging from a hose nozzle, one would have to wait long enough for the necessary amount to make its way out of the nozzle and into the bucket before building up the level of energy an electron needs to absorb before it can finally escape. This point would be reached for a typical light source something like a minute after the light is switched on. But since light comes instead in highly localized bursts of energy (more like rocks than like water), interspersed by relatively large regions of empty space, the very first rocks to be thrown can find their mark, and the photoelectric current appears almost instantly. This means that light must consist of *particles*; it simply would not happen this way if it were a wave phenomenon.

Einstein summed it all up with the simple equation

$$E_{max} = h\nu - W \, , \tag{14-1}$$

where E_{max} is the energy of the most energetic electrons, those emitted from the very surface of the metal, and is directly proportional to V_{max}, the cutoff voltage which can stop these electrons. (It can in fact be shown that $E_{max} = eV_{max}$, where e is the electric charge of the electron, 1.6×10^{-19} coulombs.)

The universal quantity h is of course Planck's constant; W is the so-called *work function* of the metal, namely, the energy required for an electron to escape from the very surface. Thus equation (14-1) states that the energy E_{max} of the most energetic electrons, those ejected at the surface, is equal to the energy $h\nu$ of the absorbed light quantum, *minus* the energy W such electrons lose in escaping from the surface.

If we plot E_{max} for a given metal against frequency ν, we obtain the algebraic equation of a straight line, like

$$y = mx + b \, , \tag{14-2}$$

where b (in our case $-W$) is the point at which the plotted line crosses the y-axis, and m (in our case h) is the slope of the line. Millikan obtained such a graph in 1916 by measuring V_{max} at various frequencies and plotting the corresponding E_{max} vs. ν. Not only was the equation that of a straight line, as Einstein had predicted, but the constant h, which Planck had determined by matching his theory of blackbody radiation to the observed data, agreed with the h of the photoelectric effect to within less than half a percent!

Actually, the story does not end here. Many physicists for a long time remained unwilling to accept Einstein's quantum theory of the photoelectric effect in spite of the fact that they could think of no other explanation. It made no sense in the framework of classical electricity and magnetism in view of Young's experiment; like Planck's quantized oscillators, it appeared to be some sort of fluke which just happened to give the right answers. A concept as difficult to accept as the quantum theory would necessarily require additional confirmation. But this was not long in coming. It is called the *Compton effect*, and was discovered in 1923.

In order to appreciate the Compton effect it is necessary to understand the simple physics which goes into the playing of billiards or marbles. When a ball strikes another ball *head on* it gives up the greatest possible amount of its original energy to the stationary ball, as in the billiards game of Fig. 14–3a. The previously stationary ball then proceeds in the same direction as the moving ball had before the collision. However, if the stationary ball receives only a *glancing* blow, as in Fig. 14–3b, it absorbs relatively less energy, much of it still being retained by the moving ball; however, the struck ball now moves not in the original direction but off to one side, while the striking ball moves to the opposite side.

Compton reasoned that if Einstein were correct in his quantum theory of light, it ought to be possible to play billiards with light quanta of energy $h\nu$. A fast moving ball would correspond to a large value of $h\nu$, i.e., a high frequency of light, more toward the blue or violet end of the spectrum. A slow moving ball would have a smaller value of $h\nu$, more toward the red end of the spectrum. The target of these light quanta was to be a relatively stationary electron. Detectors placed at various angles from the direction of the incident light would provide "pockets" to catch the light quanta and electrons rebounding in the various directions; and if these "pockets" also measured the energies of the particles, i.e., the color of the light and the kinetic energy of the rebounding electrons, it should be possible to make an excellent model of a billiards game.

Actually, Compton used X-rays, which have much higher frequency (shorter wavelength) than visible light (less than an Angstrom, whereas visible light is measured in thousands of Angstroms).

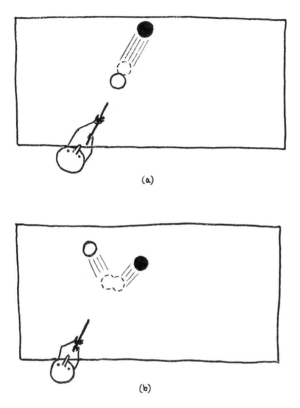

Fig. 14-3 *"Compton Effect" with billiard balls. (a) In head on collision the struck ball absorbs the most energy. (b) But in glancing collision the struck ball absorbs less energy and is deflected to one side, while the striking ball is deflected to the other.*

Poet On the one hand you are now claiming that light is a particle; on the other you still speak of "frequency" and "wavelength," terms which you have used to describe waves. Why this mixed semantics?

Young's experiment still sticks in our throats. It is the skeleton in the closet. The wave theory is a part of our scientific heritage, and we cannot discard it like an old shoe. Young's was a good experiment; there was nothing wrong with it. It has merely been supplemented with new information. Thus even while we play ball with the newly acquired particles we continue to address them in the old wave language.

The advantage of X-rays over visible light is that their energy $h\nu$ is considerably greater than the *binding energy* which ties each electron to its respective atom. We should like our target ball to be free and unencumbered; otherwise the game will be cluttered by the presence of a lot of other balls. The very energetic X-ray quantum hits the electron hard enough to snap the "string" which ties it to the rest of the atom; it is as if the other particles were not even there.

Compton's target was a thin piece of metal foil through which he passed the incident X-rays. The process is modeled in Fig. 14–4a. As the (negatively charged) electrons scatter in various directions, the *photons* (which is what these quanta came to be called) rebound, but with *less energy* (longer wavelength) than they had to begin with, since they have imparted some of their energy to the struck electrons. The lower energy of the scattered photon means a smaller value of $h\nu$, hence lower frequency ν and from equation (13–2) therefore greater wavelength λ; it does *not* mean that the light is less intense, as we discovered when we analyzed the photoelectric effect. Figure 14–4b reproduces Compton's curves as they were published in the *Physical Review* of 1923. Each of the four curves corresponds to light scattered at different angles θ measured from the incident direction, i.e., different "pockets" of the billiard table. Of course many collisions are represented by any one curve. The intensity of the scattered light (the number of photons) received in each of the four "pockets" is plotted in the vertical direction vs. wavelength of light in the horizontal. The incident light starts out at wavelength λ_0, and the scattered light has longer wavelength λ_1 (lower frequency and hence lower energy). Both wavelengths appear as intensity peaks in the curves. The loss in energy of the scattered photon, as evinced by its greater wavelength, is what we should expect in a given scattering direction for the billiard ball model of Fig. 14–3. Other experimenters also caught the recoiling electrons, correlating them in time with the scattered photons and measuring their energies. They found that the pieces all fit together. The photons and electrons had energies and were moving in precisely those directions before and after the collisions which they would if they had been billiard balls. Compton had indeed played ball with light particles! Einstein's quantum theory of light had been confirmed. If you still have any doubts, try playing billiards with water waves; this should help to convince you that light has to consist of particles.

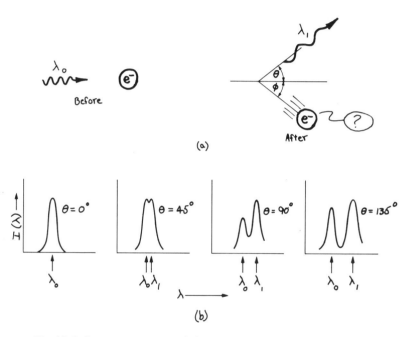

Fig. 14–4 *Compton scattering of photons against electrons. (a) The scattered photon has longer wavelength (less energy) than the incident photon. (b) Compton's original results, plotting intensity vs. wavelength of scattered light, as published in* Phys. Rev., 22, 409 (1923).

Poet Why are the bumps so wide, and why are there *two* of them in all but one of Compton's curves?

The width of the peaks is due to the fact that when we say we are using light of a particular wavelength and energy we mean *mostly* that wavelength; we do not obtain perfect filtering, and there-

fore closely adjacent wavelengths are always present. The first
graph, $\theta = 0°$, corresponds to *"forward scattering"* events in which
the photon essentially missed the electron, hence coming through
with its initial energy and its original wavelength λ_0. At scattering
angle $\theta = 45°$, the electron was struck a "glancing blow" and the
photon lost some of its energy, emerging now with a slightly longer
wavelength λ_1. The larger scattering angles of 90° and 135° approach the head-on collision condition $\theta = 180°$ (not shown because
the original source of light would get in the way of the "pocket");
the scattered photon loses progressively more of its energy, resulting in greater increases in wavelength.

However, there is an additional surprise: At all angles light is
also to be found scattered at its original wavelength λ_0; that is the
second "bump" or peak. If light consisted of waves instead of
particles, this λ_0 peak is all that would be present in every case.
In the wave picture of light we think of electromagnetic waves of
frequency ν_0 setting the electron into oscillation, also at frequency
ν_0, just as they did the hammers that went up and down in the "people wave" of Chapter 12. As a result this oscillating electron in turn
radiates electromagnetic waves in all directions θ at the original
frequency and hence the same wavelength λ_0.

Poet But presumably the wave picture has now been disproved.
Why then do we see the other peak?

This scattering of light at its original wavelength is no doubt the
spirit of Thomas Young reaching from the grave to restore perspective by reminding us that light was once a wave and shall always be
so! The phenomenon is known as *Thomson scattering* (no pun
intended), and has been analyzed by classical wave theory. However, the latter fails to account for the longer wave peaks λ_1 which
only the quantum theory predicts. On the other hand, Einstein's
quantum picture does not account for the presence of the λ_0 peaks.
These are strictly a classical wave phenomenon, and correspond to
the fact that X-rays can also produce wave diffraction effects such
as Young obtained with visible light. Thus in Young's experiment
light acts like a wave, in the photoelectric effect it behaves like a
particle, and in Compton scattering it displays both properties simultaneously. We shall discover how the new science of quantum
mechanics resolved all these conflicts and paradoxes.

First, however, we take note of a mathematical result which will be useful to us in what follows. Collisions of billiard balls obey two important conservation laws. The first is called energy conservation, and we have already encountered this principle. The energy in this case includes merely the kinetic energy of motion of the balls, and its conservation in a collision of two balls may be stated as

$$(E_1 + E_2)_{\text{before collision}} = (E_1 + E_2)_{\text{after collision}} , \qquad (14\text{--}3)$$

where

$$E_1 = \tfrac{1}{2}m_1 v_1^2 \qquad (14\text{--}4)$$

is the kinetic energy of ball 1 in terms of its mass m_1 and velocity v_1, and E_2 is a similar expression for ball 2. Some of the energy of ball 1 *before* collision may be transferred to ball 2 *after* collision (or vice versa), and this transfer of energy appears in the form of changes in the velocities of the two balls. If we know the velocities v_1 and v_2 before collision, their velocities after collision must satisfy equation (14–3). (Of course, v_2 is usually zero before the collision, and in conventional billiards $m_1 = m_2$, but we could design a more general game in which the balls have different masses.)

Unfortunately equation (14–3) is not enough to predict what happens as the result of a collision, since there are two unknowns, namely, the velocities of the two balls after the collision. Another equation is therefore required to solve the problem; this second law is called *momentum conservation*. It takes note of the fact that if something was moving to the right before the collision, something had better also end up moving to the right *after* the collision; this "something" which is conserved is called the momentum, and for a moving billiard ball it is defined as the product of its mass and velocity. Thus the momentum (designated by the symbol p) of the ball 1 is given by

$$p_1 = m_1 v_1 , \qquad (14\text{--}5)$$

and there is a similar expression p_2 for ball 2.

If ball 1 is moving in a given direction (say, to the right), we can take account of this by designating v_1 and hence p_1 as positive quantities; for the opposite direction (to the left) we then consider

v_1 and p_1 to be negative. Similarly, if there is another dimension to the problem, like up and down, we may consistently label "up" positive and "down" negative.

Conservation of momentum requires that

$$(p_1 + p_2)_{\text{before collision}} = (p_1 + p_2)_{\text{after collision}} , \qquad (14\text{--}6)$$

where p_1 and p_2 before and after collision may be positive or negative or zero. Equations (14–3) and (14–6) now make it possible to determine v_1 and v_2 after collision if we know what they were initially. Of course, if a ball happens to be moving in neither a strictly up-down or right-left direction, we may still apply equation (14–6) to the right-left portion of the motion by considering the *components* of velocity and momentum in the right-left direction; likewise we can write this equation again for components in the up-down direction alone. Thus equations (14–3) and (14–6) always enable us to predict the result of a collision when we know the velocities before collision. Furthermore, these equations may be extended to any number of balls (or series of collisions), always enabling us to predict the outcome.

If you have played billiards you have had occasion to solve such a system of equations, although typically on a very intuitive level, and without giving expression to words like "energy" and "momentum." Some players of course just close their eyes and their minds and hope for the best; but Compton certainly permitted himself no such luxury when he verified that photons were exhibiting momenta and energies corresponding to those of rapidly moving billiard balls.

Poet What happens if there is only one ball, and it collides with the edge of the table? Its velocity and momentum to the right now become velocity and momentum to the left. Why isn't this change from positive to negative momentum a violation of momentum conservation?

When a ball rebounds from the edge of the table, the table plays the role of the "other ball." Not only does the table rebound with equal and opposite momentum, but the room, the house, and even the entire earth rebound with it. It is just that this composite mass is so large compared to that of the ball that its velocity is pretty negligible. However, the product $p = mv$ is *not* negligible for such large m and small v, but is of precisely the amount necessary to account for the change in momentum of the ball when it reverses itself and hence changes sign.

This diversion into the mechanics of billiards has had the purpose of introducing the reader to the concept of momentum. It turns out that the momentum of a photon, i.e., a particle of light, is an important clue to the solution of our wave-particle riddle. The photon is certainly not like other particles. It travels at the speed of light (since it *is* light), and hence represents a rather fast game of billiards. We have already seen that Einstein in accounting for the photoelectric effect concluded that light comes in particles of energy $h\nu$, where ν is the frequency which Young discovered the light to have when it was a wave. Thus in saying that the energy is given by

$$E = h\nu , \qquad\qquad (14\text{--}7)$$

we are in one breath speaking of it as both a particle and a wave. The energy E (if it is a particle) is apparently related to the frequency ν (if it is a wave) by Planck's constant h. But if we consider it to be a particle, then what is its mass? Physicists have made determinations of the mass of other particles; thus the electron mass has been measured as 9.1×10^{-28} gram. What is the mass of the photon? The answer to this question appears in our mass-energy conversion *Gedankenexperiment* of Chapter 9. If we continue the line of reasoning developed there, we find the mass of the photon to be precisely zero! (The proof is left to the Appendix.) It may also be shown that the momentum of such a relativistic massless particle, i.e., the quantity which is conserved in collisions, is not given by equation (14–5) at all; expressions like (14–4) and (14–5) are actually only approximations deduced in effect by people observing conventional billiards games before they knew about relativity, and apply only to particles moving at speeds much less than that of light. The correct relation (see Appendix) between energy and momentum of a massless particle moving at the speed of light turns out to be simply

$$E = pc . \qquad\qquad (14\text{--}8)$$

Thus the energy of such a particle is the product of its momentum p and the speed of light c. This detail was most important to Compton in the analysis of his high-speed ball game.

When we combine equations (14–7) and (14–8), the momentum of a photon is seen to be

$$p = h\nu/c \ . \tag{14-9}$$

But we recognize the quantity ν/c to be simply the reciprocal of the wavelength λ in equation (13–2).

Hence the momentum p of the photon is related by Planck's constant h to the wavelength λ of the light by

$$p = h/\lambda \ . \tag{14-10}$$

This most important conclusion, relating the wave aspect of light to its particle aspect, is the object of having made the reader go through so much theoretical development. Perhaps after he has come to appreciate the significance of this harmless-looking expression he will then decide whether to accept it on faith or explore it in greater depth.

But now we return to our paradox. Could light have been a particle in the eighteenth century, turned into a wave in the nineteenth, and converted back into a particle again in the twentieth? Or is it more accurate to say that it was a wave for Young, and a particle for Newton and Einstein? Or is it perhaps a wave on Mondays, Wednesdays, and Fridays, a particle on Tuesdays, Thursdays, and Saturdays, and both or neither on Sundays? The surprising thing, as we are about to discover, it that there is more truth than jest in all these statements.

Poet Perhaps this is more of a paradox to a physicist than to a poet. Why should anyone be upset over the fact that two things are happening at once? One is making waves and the other particles —like splashing water and throwing rocks at the same time. You appear to have discovered an equation which somehow relates the light wave to the light particle.

This is probably the best assessment of the situation one could make in light of the evidence so far presented. It is not a very satisfying explanation, however. The fact is that light in all these experiments is manifesting itself strictly as an electromagnetic disturbance. It is not a case of two different phenomena, like the splashing of water and throwing of stones. A signal travels to us across large regions of space without knowing which experiment we are going to perform on it; but when we then look at it one way we find it to be arriving continuously, and when we look at it the other it is instead coming in compact bursts interspersed by relatively large space containing absolutely nothing. If we ask about interference, nature answers in the wave language; if we try to increase the current in a photoelectric cell, it replies in the form of particles. Furthermore, this happens even when we do both experiments with the same light. Always the answer as to what constitutes the truth seems to depend upon what questions we ask.

Einstein must have worried about this dual character of light, but he evidently expected that a more or less conventional resolution of the paradox would eventually be found. Like the reader he considered it a peculiar anomaly of the electromagnetic phenomenon; what he did not know in those early years of the quantum theory was that it is not light alone which behaves in this strange manner, but all of nature, and that the old pictures and categorizations which had sufficed until this century were inadequate to cope with the new problems.

People were still wedded in those days to the comfortable belief in their own objectivity. Truth was always presented sooner or later in neat clean packages by a solicitous providence. In science as in human relations everything was sharply defined. There were waves and there were particles, there was right and there was wrong, reality and illusion, art and artlessness, freedom-loving democracies and imperialist dictatorships. In plays and motion pictures one could tell the "good guys" from the "bad guys." Children respected their elders, minded their manners, saluted the flag, took oaths of allegiance, worshipped their gods, and knew who they were. People either obeyed their governments in peace and war or overthrew and replaced them with systems they considered more appropriate to the changing times. But always there was a rigid structure, even when it brought a new way of life, with new doctrines, new ideals, and new slogans to replace the old. In many ways it was a good time to be alive, for people knew all the answers, even if they were different. They had not yet discovered that the answer depends upon the question, and that reality sometimes has many faces and speaks with many voices.

It is possible that some of the present confusion about the state of the world is related to an increasing awareness that the problems of the times do not fit the dogmas and fragmentations of the past, and that the old pictures which used to make everything so clear do not adequately describe the current scene. In the following imagined dialogue between a Russian and an American, for example, the subject is freedom, but the answers depend on what questions are being asked.

American Our society is more free than any other. It has always served as a bastion of freedom throughout the world.

Russian Come now. Even your news media are embarrassed to use such clichés nowadays.

A Our news media speak their minds, unlike yours.

R Your media speak for the capitalist class, certainly not for the workers. Of what value is such freedom to the people in your ghettos?

A Have you ever seen an American newspaper? Our editorials run the gamut of the political spectrum from left to right. How can you compare this to your country, where only one voice is heard?

R But that voice belongs to the working people.

A When you say people you mean the government.

R In a classless society it becomes possible for the government to be indistinguishable from the people. Since the means of production are owned in common, there is no longer a class struggle; hence there are no basic conflicts.

A Is this why dissent is never permitted?

R Naturally we do not allow Fascists and warmongers, as you do. Freedom is not served by giving a voice to those who want to destroy it.

A Everybody claims the other fellow is against freedom. The operational question is whether conflicting opinions are heard. Our system passes this test, and yours doesn't.

R No, the operational question is whether the broad masses of people are represented by the government, or whether the country's economy is based on foreign wars, and ethnic minorities are denied their full share of the wealth and social structure. War and poverty are the problems of our time, and they constitute the operational test of freedom. Your conflicting opinions are merely the sounds of rich people arguing among themselves.

A But there is dissent. In your system the silence is deafening.

R An oligarchy always allows freedom for the ruling class— freedom to exploit other people.

A What you call our ruling class is chosen in a free election.

R Your free elections are like your news media; they are paid for by the wealthy and operated by the professionals.

A It is possible that a wealthy man may win more votes in an election than a poor one, but the voters have a choice, which is hardly the case in your system.

R This choice between two virtually indistinguishable parties is only a facade, a Potemkin village, like the token integration of your companies who carefully maintain a small quota of black employees. The Democratic convention of 1968 showed what happens when the chips are down. In the primaries the people voted against the war whenever they had the opportunity; but no antiwar candidate was permitted to run by either party on election day. It is what is called in your language Tweedledum and Tweedledee, and in ours Dobchinsky–Bobchinsky.

A We do not claim to have a perfect system. But there is machinery available for correcting mistakes. Anyhow, I think we are getting off the subject.

R All right, I'll change the subject. Take the case of the blacks. Surely you won't claim that a child in Harlem today is born into freedom.

A These economic and racial inequities are results of our past history. We are doing everything to correct them. It is simply not possible for a foreigner to appreciate the difficulty. When you consider how much has been done in a short time, and that black people were slaves only a few generations ago, it becomes obvious that enormous progress has been made. In any case we were discussing freedom of expression, not equality of opportunity. Whatever they may lack, black Americans are free.

R   Free to live where they wish, for example? Free from having to participate in colonial wars? Free to have the best medical care? Do the black people consider themselves free? Have you asked them lately?

A   At least they are not herded into slave labor camps or executed at the whim of a police state.

R   I admit we made some mistakes in the past. A foreigner cannot understand what it means to live in a socialist country surrounded by enemies and constantly under the threat of invasion. Sometimes compromises are necessary in order to secure freedom in the long run. Anyhow, the cult of personality is no longer an issue in our country. So why rake over old coals? Now *you* are getting off the subject.

A   All right. If the fact that your press is controlled, your literature censored, and your artists confined to "socialist realism" is irrelevant, whereas ghettos and poverty come under the subject of freedom, then it must be that standard of living is a criterion. Our "oppressed workers" are the richest in the world; even in Harlem there are more cars and transistor radios than in Moscow.

R   We have raised the standard of living at a phenomenal rate when you consider the feudal origin of our country in the first World War and the devastation it suffered in the second. But what have cars and transistors to do with freedom? You simply do not understand the question.

questions for the reader

1. Consider an atom to be a sphere of radius 10^{-10} meter. The energy required to free an electron from the atom is known to be of the order of 10^{-19} joule, and this is consistent with the actual cutoff voltage V_{max} observed in the photoelectric effect. Calculate how long it would take after switching on a one-watt

(one joule per second) light source at a distance of one meter for this quantity of energy to fall upon an atom. This is how long one would have to wait for the photoelectric current to begin if light were a continuous flow of energy.

2. A billiard ball moving at velocity *v* strikes head-on another ball of equal mass which was at rest. Prove that after the collision there will again be one ball moving at the original velocity *v*, and another ball at rest.

3. A billiard ball or electron loses momentum by having its velocity reduced. But a photon always moves at the speed of light; what therefore changes when a photon loses momentum?

4. If you discovered a new civilization on a strange planet, what questions would you ask in order to determine whether it is a free or a slave society?

15   A particle is a wave

*Everything has two handles—one
by which it can be borne; another
by which it cannot.*

EPICTETUS
Enchiridion

The origins of the quantum theory are to be found in the work of Planck and Einstein. But it has been said of Planck that he did not fully appreciate the significance of what he had done; and Einstein became increasingly uncomfortable with the philosophical conclusions which emerged from the physics. New names appeared on the scene—names such as Bohr, de Broglie, Heisenberg, Schrödinger, Pauli, Born, Dirac. Some of them, like the youthful Heisenberg, had little sympathy with the doubts and hesitations of the older physicists in accepting the interpretations of the atomic experiments and the strange view of nature which they presented. Einstein, who in his youth had been the stormy petrel of the physics establishment, became the conservative elder statesman who challenged and questioned the new concepts as skillfully as he had once propounded his own. The counterarguments he offered and the paradoxes he posed belong to the literature of the period. In this he was performing an important function; for science thrives on criticism, and it is only in the cold light of skepticism that the truth may ever be found.

To the day of his death Einstein remained unreconciled to what he regarded as the view that nature is a game of dice God plays with the universe. Yet he never succeeded in finding a satisfactory alternative to quantum mechanics, and its results and predictions have survived nearly half a century of rigorous experimentation and technological achievement.

At this point we abandon the historical approach and take a great leap forward. Instead of recounting all the observations comprising the evidence, we turn our attention to one most impressive experiment, which we consider conceptually rather than literally. In order that the reader may experience the full impact of the events as they descended upon the physics community, we remove it from the confines of the laboratory and transform it symbolically to a more familiar environment. Subsequently we shall take leave of our parable and return to some of the actual experiments.

You have persuaded the owner of a bowling alley to make some minor changes in his floor plan; the objective is the pursuit of knowledge, and also you want to try out a new game you have just heard about. The alleys themselves are removed and replaced by a flat level floor. A vertical plywood partition is then constructed as shown in the plan view of Fig. 15–1a. This barrier has two holes, or passages, which will permit bowling balls to pass through. A number of your friends are stationed in a row at one end of the room and provided with an abundant supply of balls. The lights are turned out, and with the room in darkness the players proceed to roll their balls at uniform speed and in the same direction as if they were trying to make a strike in a conventional alley without a vertical partition. Of course now most of the balls are stopped by the barrier (and automatically removed from the scene like expended bowling balls). However, some of them manage to pass through the two spaces in the plywood and make their way to the other end of the room, where they strike the far wall (called the detector) and are likewise removed.

Now this wall which is the ultimate target of the balls has been covered with special paper which retains a mark every time it is struck by a ball. After an hour or so, when everyone has grown tired of the game, the lights are turned back on, and all go up to have a look at the marks left on the wallpaper of the detector by balls which passed through the holes in the barrier.

What you expect to find is two bunches of marks, one behind one hole and the other behind the other, as in Fig. 15–1b. But to

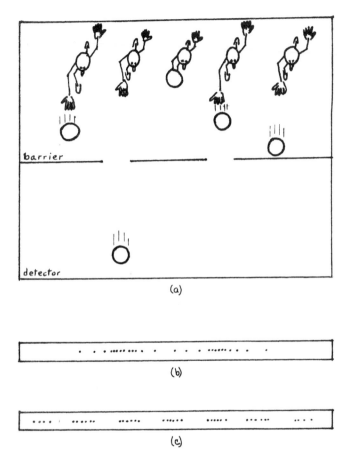

(a)

(b)

(c)

Fig. 15-1 *The parable of the bowling balls. (a) The modified bowling alley. (b) Expected pattern of hits. (c) Observed pattern of hits: interference of particles.*

your surprise you discover instead the unusual-looking pattern of Fig. 15–1c. Toward the center there is a group of hits, but then to either side there appears a region with no hits whatever. This is followed (again on either side) by another set of clustered hits, and then once more a dead space. Such alternating *fringes* occur periodically; in fact they bear a striking resemblance to the *maxima* and *minima*, or interference fringes of Fig. 13–4, which we observed first in the case of water waves, and later with coherent light passing through two slits in a barrier. This is a bit puzzling, for what do bowling balls have to do with water waves and light sources? Why should real particles experience effects characteristic of waves?

Someone suggests blocking one of the holes. The marking paper on the detector wall is replaced by a clean sheet, and the game is resumed. First the left hole is blocked and the game played for an hour, after which this hole is uncovered and the right hole blocked for the second hour. Of course a lot more balls get wasted, but this is no great loss to an affluent society. In any case *alternate* holes used one at a time over a two-hour period ought to produce the same effect at the detector as *two* holes exposed for one hour.

To everyone's surprise this is not the case at all; with one hole at a time the interference pattern fails to appear, and is replaced instead by that of Fig. 15–1b, which we expected to see in the first place. But when the experiment is repeated with two holes it again generates the interference pattern. For some inexplicable reason balls in the one-hole experiment behave "normally," whereas the two-hole balls act more like waves. In other words, a ball passing through one hole is affected by the presence or absence of the other hole! Otherwise why should there be a different pattern in the two cases?

Furthermore, the dead spaces, or no-hit regions of Fig. 15–1c, which appear when *both* holes are open, occur in places which actually receive hits when only *one* hole is open. Why should opening a second hole *prevent* balls from arriving at portions of the detector which are accessible with a single hole?

It is proposed that perhaps balls bouncing off the edge of a hole collide and interfere with balls from the other hole. This could explain the pattern of hits. There is a way to test this theory. We instruct the players to roll their balls less often; in fact we slow down the process to such an extent that a particular player may roll a ball only once in ten minutes, or even once a day, or once a week. It is certainly possible to slow things down to the point where the possibility of interference between different balls is completely eliminated.

Now we have finally perfected the game; it should clearly make no difference whether one hole is uncovered or two, provided of course the game lasts long enough to obtain a large number of hits.

But something very disturbing happens. The "slowdown" makes no difference whatever! *With both holes* the interference pattern is produced, and *with one hole* at a time it is the other picture, just as before. The "interference" of balls passing two holes is rather hard to reconcile with the fact that it is now virtually impossible for more than one ball to be in play at a time. How can a single ball interfere with itself?

If you are completely honest and willing to divest yourself of all prejudice, you will have to conclude that a single ball is passing through two holes in the partition. Either it goes through both of them simultaneously, or in going through one hole it puts out some sort of feeler to sense whether the other hole is open or closed; or else after clearing one hole it comes back around and then goes through the second hole.

Everyone in the bowling alley has by this time begun to feel very insecure. The game (which is no longer a game) is repeated many times; but the results are always the same. "Why are we doing this in the dark?" someone asks. "I want to *see* the ball pass through two holes." No one seems to remember why the lights were turned out; it is apparently part of the mystique of the game. This requirement is now verified; when the game is played again, this time with the lights on, there is in fact no interference pattern.

Another suggestion: Observers are stationed in the dark alongside either hole (on the opposite side of the barrier, so as not to be struck by oncoming balls). When one of them hears a ball coming he briefly flashes on a faint light, just enough to show whether it is passing through his hole. If both observers call out, "There it goes!" simultaneously, you will report to an eagerly awaiting world that a ball has been discovered in the act of being in two different places at the same time, a feat which until now has been possible only with waves, never with particles.

But no such coincidence ever occurs; the ball passes through one hole, or the other hole, but never both holes. However, when the pattern of hits at the detector is examined the interference phenomenon is again found to have vanished. A two-hole game played without observers produces interference; but when observers are stationed to catch the dual occurrence *in flagrante delicto,* as it were, the interference pattern disappears!

Can there be something about the presence of the observers or their tiny flashlights which inhibits the balls or the players? It would appear so. The observers are accordingly removed, and the game resumed in total darkness. However, all unbeknown to the players, this time you have covered every ball with a thin layer of chalk or marking dye which leaves a trail on the floor. The instant a ball is heard to strike the far wall you stop the game, turn on the lights, and rush forward to examine the track left by this particular ball. There is a track, all right; but it passes through only one hole. The process is repeated until a pattern of hits has again been recorded on the marking paper at the detector. The pattern turns out to be that of Fig. 15–1b; interference fringes are nowhere to be seen.

Let us try to summarize the results of our experiment, as the physicists summarized theirs: Particles passing a multiple-hole barrier are subjected to what appears to be a diffraction or interference effect previously observed only in the case of waves encountering a similar barrier. This indicates that a particle is not necessarily localized, as we believed, but may instead be spread throughout the entire apparatus, penetrating all exposed apertures and eventually interfering with itself like a wave. However, any time we try to observe the *position* of the particle during the process, it immediately reverts to its classic role, becomes localized, and is no longer susceptible to such interference. We have not yet managed to understand why this is so, or what it is that is "waving," or why no one has ever noticed it before, but apparently this is the way things are in the world.

It is appropriate to warn the reader against rushing over to the nearest bowling alley and acting out this parable. If it were so simple the quantum-mechanical nature of the universe would have been discovered long ago. There is good reason for our failure to observe such behavior in the case of macroscopic objects like bowling balls or railroad trains or rocket ships, and this reason will shortly become apparent.

Developments in physics are a seesaw process in which experimentalists and theoreticians alternate; an experiment to test a theory produces an inconsistency, the theory is reexamined and revised to resolve the paradox, and the predictions of the modified theory are then tested in further experiments. Occasionally a small crack in the wall of understanding develops into a large fissure, and a radical reformulation of fundamental concepts becomes necessary; we saw this following the Michelson–Morley experiment, and our bowling ball parable suggests another such cataclysm in the offing. However, on rare occasions someone jumps the track and anticipates coming events with a *Gestalt*-like flash of insight. It is not always clear whether such individuals are truly inspired or just lucky; they appear somehow to step out of time in seeing things a moment before they happen. If the physical discovery is already in the air it is not likely that such vision is entirely accidental.

There was once a graduate student named de Broglie who had an amusing thought. If light, he said, which we know to be a wave, can under certain circumstances (as in the photoelectric effect) assume the aspect of a particle, then perhaps what we are in the habit of thinking of as particles can likewise behave as if they were waves. It is the old symmetry argument. Since physicists, unlike pure mathematicians, prefer their conjectures to be based on what has been observed, there was some doubt whether to accept de Broglie's doctoral dissertation, which explored such a whimsy in all seriousness. It was 1924 and the experimental evidence for the diffraction of particles had not yet been discovered; but it was about to be. De Broglie's suggestion gave a sharp impetus to the development of quantum mechanics; yet if it had not been for the successful experiments which followed, his thought would have remained an entertaining diversion, like the beautiful cadenza a violinist is allowed in the course of playing his concerto.

De Broglie argued that the particle aspect of light, or electromagnetic radiation, is manifested in the interaction of the radiation with matter, as with the photocathode in the photoelectric effect, or the electrons in Compton scattering. On the other hand the wave aspect appears in the propagation of the radiation as it makes its way around obstacles and through apertures. Thus light has two "handles"; in asking how it interacts with matter we are grasping one "handle," and in asking how it propagates, the other. It would be naive to insist that either of these represents the "true" nature of light, and that the other is merely an illusion; it just happens that we had occasion to discover one before the other. But this is no reason for maintaining a closed mind.

Now if it should likewise happen that objects we normally think of as particles have a wave side to them, it would again be that we have simply not had any occasion to observe it. How would the wave nature of particles be manifested? We take our cue from the experience with light. The propagating electromagnetic waves *point the way*, telling the particles where to go, and when the photons finally make their appearance it is always in those places which were able to be penetrated by the waves.

Thus objects like electrons and bowling balls and perhaps even people would likewise be guided or borne by what de Broglie called *pilot waves*, which point the way, telling the particles where they are most likely to be found. It was all very mysterious and metaphysical-sounding, and hardly in the nature of the business of people who are supposed to be "midwives of technological progress." But of course it was only a proposition. No one was pontifically insisting that it had to be so; it was just a suggestion for future experiments.

Now if particles were waves, what would be their wavelength? Again we take our cue from the case of light. The wavelength λ of the light was found in the last chapter to be related to the momentum p of the photons by equation (14–10), which we now rewrite as

$$\lambda = h/p . \tag{15-1}$$

Therefore all we have to do to find the wavelength associated with a moving billiard ball is divide Planck's constant h by the momentum p of the ball, where the momentum defined in equation (14–5) is the product of the mass and the velocity of the ball. The wavelength given by equation (15–1), known as the *de Broglie wavelength* of the particle, tells us what kind of wave to look for. We still do not know what it is that would have to be waving, but since we know its wavelength at least we have a clue to the conditions under which it may be observed.

The wave nature of light was discovered in Young's experiment; to establish the wave nature of particles we need another Young's experiment. However, if the interference fringes are to be seen at all, they will have to be visible to us, their human observers. Let us therefore determine some of the relevant dimensions. Equation (13-1) gives the distance from the central maximum to the first minimum of an interference pattern; it therefore is a rough measure of the width of a fringe. In this equation the ratio $\lambda/2d$ (half the wavelength divided by the distance between slits) is multiplied by f, the distance from the barrier to the screen. If the fringe proves to be of such order as to be visible (not too large and not too small), then we have a chance of discovering the *pilot waves*.

Let us therefore compute the de Broglie wavelength of, say, a bowling ball. If the mass of the ball is taken as one kilogram, and its velocity one meter per second, the momentum is

$p = mv = 1$ kg m/sec.

Dividing this into Planck's constant $h = 6.6 \times 10^{-34}$ joule-seconds gives us a wavelength

$\lambda = h/p = 6.6 \times 10^{-34}$ meter.

In view of the 34 zeros to the right of the decimal point in this number, it is indeed a submicroscopic wavelength, compared even to typical wavelengths of light, which run about 5000 Angstroms, or 5×10^{-7} meter. Thus the wavelength of visible light is some 10^{27} times as great as that of a bowling ball. If we separate our holes or slits in the barrier by a distance $d = 1$ meter, and make the distance f to the detector, or screen, 10 meters, the width of a fringe should become, from equation (13-1),

$$D_1 = \frac{\lambda f}{2d} = 3.3 \times 10^{-33} \text{ meter.}$$

One can appreciate how small a distance this is by comparing it to the diameter of an atom, something like 10^{-10} meter, or the nucleus of an atom, 10^{-15} meter; thus our interference fringes would be 10^{18} times *smaller* than the nucleus of an atom. Since an atomic nucleus has never been seen even with the most powerful microscope, what chance have we of seeing fringes which are 10^{18} times smaller? Clearly it is no coincidence that quantum mechanics was not discovered by the proprietor of a bowling alley!

Even increasing the distance f to the detector from 10 meters to 100 meters, and reducing the spacing d between slits from 1 meter to 1/10 meter would still remove only two of the 33 zeros in

10^{-33} meter. We need an improvement of some 30 orders of magnitude (30 multiplications by 10) before being able to test de Broglie's hypothesis. A radical modification is required in our apparatus. The crucial factor is λ/d; this ratio has until now been much too small. Instead of a bowling ball, with its submicroscopic de Broglie wavelength of 10^{-33} meter, what we want is a particle with a wavelength not too much smaller than the distance d between slits; then we can hope to see interference fringes in the pattern of hits at the detector. Since $\lambda = h/p$, and h is a constant of nature over which we have no control, the only hope is a sharp reduction in the particle momentum p. This can be attained by reducing the mass of the particle, or its velocity, or both. Now there is little room for improvement in velocity; if we slow down the particle from its present speed of one meter per second by much more than five or ten orders of magnitude, we may grow old and die waiting for it to reach the detector. On the other hand the *mass* of an *electron* is only 10^{-30} kgm, 30 orders of magnitude smaller than a bowling ball. This could therefore solve our problem. What we need is a bowling alley in which the balls are electrons.

The Davisson–Germer experiment (1927) represents the first time creatures on this planet (and probably this planetary system) were privileged to observe the wave nature of particles. Electrons which had been collimated (so as to have a fairly uniform direction and speed) at a velocity of 4×10^6 meters per second were directed against the surface of a nickel crystal, which provided the "slits." The de Broglie wavelength of the electrons was

$$\lambda = h/p = \frac{6.6 \times 10^{-34}}{10^{-30} \times 4 \times 10^6} \approx 10^{-10} \text{ meter.}$$

In order to produce visible interference fringes it was required that the spacing d between slits be not much larger than this wavelength (thus providing a reasonable factor of λ/d); this meant that d had to be comparable to the size of an atom. It was clearly impractical to drill holes of such proportion in a plywood partition. But the crystal lattice of a polished surface represents an orderly array of atoms with spaces between them, somewhat like soldiers standing in parade formation; of course it represents a barrier not with just two "holes," but a large network of "holes." Such a pattern of obstacles and spaces is called in optics a *diffraction grating*; it produces an interference pattern when exposed to coherent light of appropriate wavelength. Thus the first observation of the diffraction of particles occurred when electrons were sent against such a barrier.

Other experimental details can be omitted here. In the original experiment the electrons were *reflected* from the surface of the crystal lattice instead of *passing through* the barrier, although in later versions penetration of a thin polycrystalline foil by the electrons was effected. What is significant is that interference fringes were observed in the pattern of "hits" at strategically placed detectors; and the width D_1 of the fringes did in fact correspond to the de Broglie wavelength λ of the electrons, the lattice spacing d between atoms, and the distance f to the detector, in accordance with equation (13–1), which had been derived for interference of light waves.

This experiment has since been repeated with many variations, including the use of other particles, like neutrons. Furthermore (and this is most significant), the same interference phenomenon persists even when the intensity of the beam is turned down to the point where the particles pass through the apparatus *one at a time!*

Thus there can be no question but that a single particle is simultaneously squeezing itself through all the available openings—a process which is normal for waves but totally unreasonable for particles in a "sane" world.

In Figs. 15–2*a* and *b* actually observed interference fringes are compared for electrons and for light, respectively. The pattern of electron "hits" has been magnified by an electron microscope; it may be seen to be the same as for light waves in an analogous experiment. In this example of the diffraction of waves and particles the phenomenon is produced with effectively only two slits. A reader interested in experimental details is referred to the original sources.

In Fig. 15–3 the phenomenon of *edge diffraction* is likewise illustrated. This is the familiar pattern of fringes which appears when coherent light passes the edge of an obstacle. Similarly electrons passing the edge of a sharp barrier produce the characteristic interference pattern *a*; an electron microscope has again been used to record the pattern of hits. The corresponding phenomenon is reproduced in *b* with light passing a similar edge.

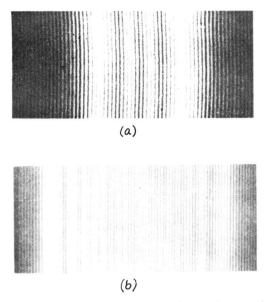

(a)

(b)

Fig. 15–2 *Comparison of double-slit interference of electrons with that of light. (a) Interference pattern of "hits" made by a coherent source of electrons which has in effect been passed through two slits, and has been greatly magnified by an electron microscope. (Reproduced from H. Düker, "Lichtstarke interferenzen mit einen Biprisma für Electronwellen,"* Zeitschrift für Naturforschung, Volume 10A, 1955.) *(b) Interference fringes developed on a photographic plate by coherent light passing through two slits. Scale has been deliberately chosen to produce results which may be compared to (a) above.* (From Valasek, Introduction to Theoretical and Experimental Optics, John Wiley and Sons, New York, 1949.)

Poet Without knowing more about it, one would suspect that you might have been mistaken in your original assumption about the objects you call electrons. If they behave like waves, perhaps that is what they always were.

One would like to be able to say that some things are particles and others are waves. Unfortunately every time we try to make such a dichotomy nature confronts us with a contradiction. If in the electron interference experiment we use as the detector a device known as a *scintillator*, which flashes every time a charged particle strikes it, there can be no doubt that the electrons are localized, i.e., that they are particles. They produce distinct flashes

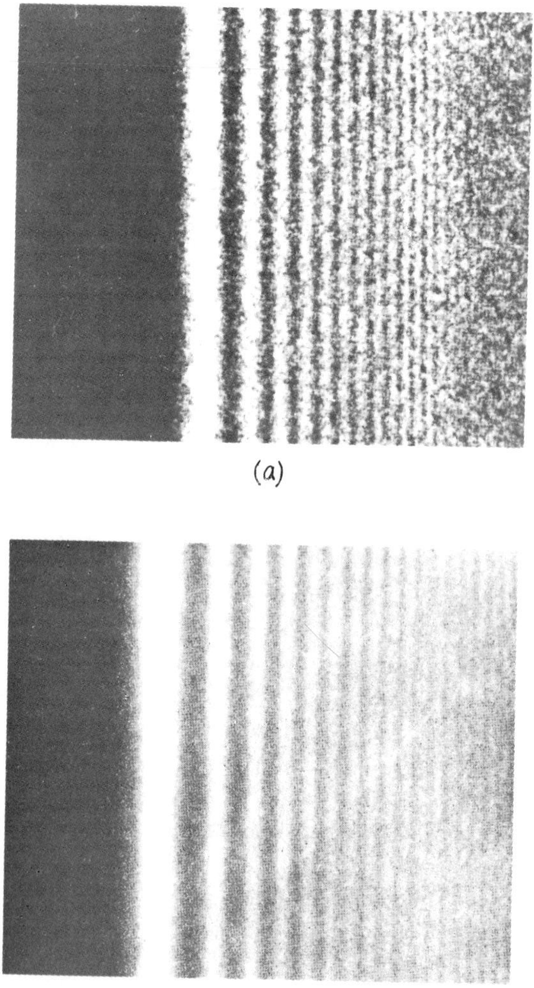

(a)

(b)

Fig. 15-3 *Edge diffraction effects for electrons and for light.* (a) *Patterns of electrons diffracted from edge of a microscopic crystal of MgO and photographed with electron microscope.* (From H. Raether, "Elektron-interferenzen," Handbuch der Physik, *Volume XXXIII, Springer-Verlag, Berlin, 1957.*) (b) *Corresponding pattern of light diffracted by the edge of an obstacle.* (From Valasek, Introduction to Theoretical and Experimental Optics, *John Wiley and Sons, New York, 1949.*)

whenever they hit, just as bowling balls make a sharp sound when they strike a wall. But the pattern of hits made by a large number of such strikes is the wave interference pattern. The individual electron arrives as a particle, but the place where it lands is determined by the behavior of a wave which has passed through the equipment. The wave makes its way through all available apertures in the barrier, like water waves in a ripple tank, but what arrives at the detector is intact as a particle.

One would wish to believe that this is a peculiar property of electrons, just as Einstein originally preferred to assume that it was a peculiar property of light. But we now observe this same *wave-particle duality* in all the particles of nature; it is required only that the mass be small enough to produce visible fringes.

Furthermore, every time we "plant" a detector in the vicinity of one or more of the slits in an effort to "catch" the particle acting like a wave and being in several places at once, it simply reverts to its particle nature, and refuses to perform; i.e., the interference pattern vanishes. This is what happens in the bowling alley parable; no matter whether we turn on all the lights or introduce only the gentlest of probes in an effort to locate the particles, it is all to no avail. When we confine ourselves to questions about the ultimate destination of the particles, they behave like waves. But when we ask where in the apparatus the particles are to be found at a particular instant, the waves are invariably "frightened away," and there is no interference phenomenon.

Physicists have been living in a world of this sort since the early decades of the present century. Most of them have by now come to accept it as a perfectly "normal" state of affairs. They even consider that they understand it. Generations of physics students have been born who know no other way of life. The reader in trying to adapt to this quantum-mechanical universe will have to do as the physicists did; he must divest himself of deep-seated prejudices and assumptions.

It is, however, worth noting that the physicists originally had to work with evidence of a more indirect nature than is presently available. It is on the atomic level that nature exhibits its most striking peculiarities; thus the earliest formulations of the quantum theory arose in attempts to deduce the structure of the atom.

An important breakthrough came in 1911. Rutherford and his collaborators had been firing beams of positively charged *alpha particles* (nuclei of helium atoms obtained from radioactive decay) at thin foils of various metals, just as we once suggested a detective might throw rocks or fire bullets at a house he has under surveillance. Rutherford discovered from the way the alpha particles scattered that there is a lot of empty space inside an atom. Most of the "bullets" were apparently passing through without encountering much of anything; but every once in a while an alpha particle bounced back as though it had made a "direct hit." The pattern of scattered alpha particles suggested that an atom consists of a relatively dense core, or nucleus, which has a positive charge, and that this nucleus contains most of the mass of the atom, even though its diameter is thousands of times smaller than that of the region occupied by the electrons of the atom.

But then what would keep the negatively charged electrons in the atom from being attracted toward the positively charged nucleus, resulting in the collapse of the atom? A solution to this problem was suggested by Bohr in 1913. He proposed that the answer was to be found in the heavens. The solar system consists of planets revolving round the sun, and these bodies do not fall inward despite the gravitational attraction of the sun. What keeps a planet in orbit is its very motion. Like a ball at the end of a string, it revolves around the sun in stable and virtually endless progression.

The Bohr model of the hydrogen atom, as first proposed in 1913, consists of a massive positively charged nucleus and a negatively charged electron rotating about it like the earth around the sun. The helium atom has two such planetary electrons, lithium three, and so on, thus building up the periodic table.

There was only one strange thing about this theory. The earth happens to be 93 million miles from the sun; but presumably this is more or less accidental. It could as easily have been 92 million, or 94 million, or perhaps even 930 million. (The last figure would not have been very good for us Earthlings, but in the grand scheme of things it might have had other compensations.)

Yet such freedom is inexplicably not permitted the atomic electrons. Only certain precisely ordained orbits are available, like rooms in a hotel; it is as if there were a law of nature forbidding a planet to take up residence anywhere between the earth and Mars.

Now the fact that the earth year has its present duration is directly related to the distance of the earth from the sun. The Venusian year is shorter, and the Martian year longer. Every planet in its (approximately) circular orbit has a characteristic speed of travel around the sun. It is precisely this speed which keeps it in its present orbit; a large burst of rocket power would cause it to swing farther outward; a loss of energy would make it fall closer inward. But at any given distance from the sun there is a unique orbital speed. Thus when someone tells you the velocity of a circular orbit, you can tell him its radius, i.e., its distance from the sun. (The actual mathematical relations appear in the Appendix.)

In order to make Bohr's atomic model fit the experimental data, it was necessary to postulate that only certain of these orbits were able to be occupied. It was like the old story of Planck's oscillators with the discrete energies, or our fable of the people on the Peculiar Planet whose heights could never come in fractions of an inch. The evidence for such *quantized* atomic orbits was the result of an experimental technique known as *atomic spectroscopy*.

When atoms are excited (as they are in the vapor of a fluorescent lamp, for example, by the passage of electric current), the atomic electrons are forced into higher energy states. In this respect they resemble an orbiting space ship which has just received a burst of rocket power. The higher we wish to make the orbit of an earth satellite, the more fuel we must expend in getting it there; by the same token electric excitation of a fluorescent gas raises its atomic electrons to higher energy states, namely, larger orbital radii.

But these electrons do not remain forever in the high new orbits appropriate to their greater energy. Instead, somewhat like an earth satellite which loses energy because of atmospheric friction, the electrons proceed to decay back down to the energy states they occupied before the excitation, i.e., back to their "ground states." In falling back into these lower energy levels the atoms are observed to emit light of a characteristic frequency (color); and the frequency ν of the emitted photons is related to the *energy dif-*

ference E between the excited state and the lower state, according
to the relationship previously suggested by Planck and utilized by
Einstein, namely,

$$E = h\nu \qquad\qquad\qquad (15\text{-}2)$$

where h is Planck's constant, and ν the frequency of the emitted
light.

Analysis of the spectrum of light emitted by the excited atoms
of a gas after such light has been passed through a spectroscope,
or prism, which breaks it into its component colors, makes it possible to compute the differences in energy of the allowed electron
orbits. Bohr was thus able to deduce the actual energies of the allowed orbits, and hence their velocities and radii. He found that
the circular orbital radius r (and the corresponding velocity v) required to keep the electron in a stable orbit is for some reason
permitted to have only those values which satisfy the equation

$$mvr = h/2\pi,\ 2h/2\pi,\ 3h/2\pi,\ 4h/2\pi,\ \ldots, \qquad (15\text{-}3)$$

where m is the mass of the electron, and h is as always Planck's
constant.

The quantity on the left of this peculiar equation (mass times
velocity times radius) is called the *angular momentum* of the electron in its circular orbit. Bohr's postulate states that nature tolerates
only those orbits whose angular momenta correspond to integral
multiples of $h/2\pi$. This was the message conveyed by the spectroscopic data.

Bohr probably had no idea why, when the planets of the solar
system can apparently assume any orbits they please, the electrons
of the atom should be restricted in this way. It was an empirical
observation, yet surprisingly accurate. Just as Planck had explained
blackbody radiation and Einstein the photoelectric effect by postulating energy *quanta* which are integral multiples of $h\nu$, Bohr now
explained the spectroscopic data by postulating discrete *quantized
values* of angular momentum for the electron orbits. This theory
of atomic structure therefore came to be called the *quantum theory*,
and the new science which eventually encompassed all the earlier
observations and successfully predicted the later ones came to be

known as *quantum mechanics*. What has all this to do with the nature of physical reality, and the fact that particles behave like waves? The reader is now in the fortunate position (unlike Bohr in 1913) of being able to appreciate the connection.

There is no *a priori* reason why a particle like an electron should jump from one allowed orbital radius to another, as indicated by the spectroscopic data, without ever occupying any of the region in between. De Broglie pointed out, however, that if an atomic electron is assumed to take the form of a wave, then there is a very plausible explanation for such behavior.

Let us rewrite equation (15–3) in terms of the momentum *p*, which we have defined as equal to *mv*:

$$pr = h/2\pi, \ 2h/2\pi, \ 3h/2\pi, \ 4h/2\pi, \ldots \qquad (15\text{–}4)$$

Since the velocity or momentum uniquely prescribes the orbital radius *r*, this equation specifies a discrete set of momenta *p*. And if one computes the de Broglie wavelength $\lambda = h/p$ for each of these momenta, the allowed orbits are just those into which one, or two, or three, or *any whole number of wavelengths* fit exactly! This is shown in the Appendix, and illustrated in Fig. 15–4 for a one-electron atom. De Broglie demonstrated that Bohr's *ad hoc* postulate becomes understandable when one tries to fit waves instead of particles into the allowed orbits. In Fig. 15–4b, for example, six wavelengths are shown fitting exactly into one of Bohr's "allowed" orbits. But in Fig. 15–4c a "disallowed" orbit is represented, in which an integral number of wavelengths does *not* fit. Thus de Broglie was clearly not being arbitrary or unmotivated.

We have now seen the wave nature of particles in two very different manifestations. In the Davisson–Germer experiment a beam of so-called "free" or "unbound" particles is found to behave like *traveling waves* passing through apertures in a barrier. In the case of Bohr's orbits, on the other hand, the electrons are in a "bound state," like a satellite tied to its parent body by a force of attraction. In the latter case the waves surrounding the nucleus are called *standing waves*; the nodes and peaks remain fixed in position, but the peaks of the waves change repeatedly from crests to troughs and back to crests again. This wave picture of an electron bound to a nucleus is called a *stationary state*.

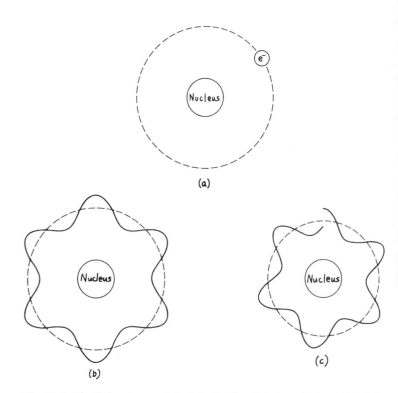

Fig. 15–4 *The Bohr atom explained by De Broglie's hypothesis. (a) Bohr's one-electron atom, with the electron as a particle revolving in a planetary orbit around the nucleus. (b) Allowed "orbit." An integral number of De Broglie wavelengths may be precisely fitted. (c) Disallowed "orbit." De Broglie wavelengths cannot be fitted at this radius.*

The important thing, however, is that material particles have in fact been shown to behave like waves. Although the Bohr atom actually turned out to be a crude first approximation, de Broglie's suggestion has been confirmed in countless other experiments involving many different types of particles. The wave nature of material objects appears to be an undeniable fact of life. What we think of as particles are also waves. Of course we actually *observe* the wave behavior only when dealing with a sufficiently small mass and hence a de Broglie wavelength large enough to detect. But since macroscopic bodies such as locomotives and space ships and people are composed of elementary particles like protons and neutrons and electrons, there is in principle no reason to disallow the wave nature even of the larger bodies.

Poet If all things consist of waves, then why have we been told that small boys do *not* cancel each other like waves when we send them to the grocer's for milk?

Scientist Who said they never cancel?

P The author said so in Chapter 13.

S As I recall it, what he actually said was *hardly ever*. The suggestion is that such cancellation would be extremely difficult to observe, like interference fringes in a bowling alley.

P Perhaps he simply likes to change with the times. At the beginning he professed a belief in an objective world. Now he isn't so sure; says it depends on the question. It is considered in bad taste for laymen to challenge currently accepted scientific theories, but the living scientists always proceed to discover the mistakes of the dead ones. Then they say, "Those other fellows were a bit confused," or, "Their equipment wasn't very good in those days, but now we know better."

S The dead scientists have not been proved "wrong." Their work has only been extended. Whenever the range of human experience is increased new principles are likely to be needed. Newtonian mechanics was not *wrong* in its analysis of bodies moving at speeds much less than that of light; in fact it is an excellent theory for the so-called "normal" range of phenomena, and we continue to use it in most engineering applications. Thomas Young was not *wrong* in his discovery of the wave nature of light; but his theory fails to account for its particle behavior. The object is to fit an orderly system to as much of nature's handiwork as one is privileged to witness. It is not a case of looking for "correct" laws; there are only better laws and worse laws, fundamental laws and *ad hoc* laws, laws which can predict future behavior and laws which cannot. The laws of nature were not designed for us and then temporarily concealed like toys hidden from a child by a playful Parent—any more than a grammar is constructed before the language itself is used. The grammatical structure is adjusted to fit the language; and when the language changes, then so must its "laws."

P   But what if one can speak the language without learning the grammar? It could turn out that your theories of grammar are too confining.

S   Even when a language is spoken fluently there is something to be learned from the study of its structure. But I do not wish to belabor this analogy. If I did, I should say that physical laws are like grammatical principles which predict the future behavior of the language, telling us not only what is happening today but also what people will be saying or thinking hundreds of years from now.

P   Unfortunately the predictions often turn out to be wrong.

S   Actually, that is not true. As science has developed its predictions have become increasingly reliable. There have been certain important oversights, but sometimes we learn even more from the mistakes than the successes. The facility of critical reasoning which you and I take for granted is the result of mistakes of the past. Before Galileo it had not occurred to people to employ the experimental method. The ancient Greeks appear to have misjudged the number of teeth in the human head because it did not occur to them to count them carefully. Much later Newton used concepts which could not be defined operationally, but as a result we have been forewarned of this pitfall. Now at last we learn that the everyday pictures used to describe the physical world are no longer adequate. In this too there is an important lesson.

We are getting ahead of our story. In the chapters which remain we have yet to interpret our "matter waves." The phenomena of interference and reinforcement have been observed with submicroscopic particles which are elementary constituents of matter; but in the case of macroscopic bodies like boys and milk containers such experiments are rather difficult to contemplate.

Of course, if we lived in a universe with a much larger value of Planck's constant h, then presumably the de Broglie wavelength

$\lambda = h/p$ would be large enough to produce visible fringes even in the movement of billiard balls and automobiles and rocket ships. But in the present universe we are left with the problem of interpreting what it all means. The apparently senseless behavior of the bowling balls as projected in our parable raises questions about the existence or meaning of the concept of objective reality; there is no doubt but that on the submicroscopic level the objective description of nature has broken down. If light is one thing in one experiment and another in the other, if an electron can be a wave when it is locked in the atom but becomes a particle when it strikes a scintillator, if a bowling ball symbolically may be either localized at a point or spread throughout the room depending only on what questions one asks about it—then it appears we must go back to the beginning and reexamine some of our fundamental assumptions. Such "restructuring" of the physical world therefore becomes our next task.

questions for the reader

1. Compute the de Broglie wavelength of a railroad car which has a mass of 10,000 kilograms and moves at a speed of 10 meters per second. Estimate how narrow the distance between two tunnels available to such a train would have to be in order for it to behave like a wave, thus explaining why railroad engineers do not know quantum mechanics.

2. An astronomer named Gulliver claims he has by spectroscopic analysis of starlight just discovered a new universe called Lilliplanck, in which Planck's constant, instead of our value of 6.6×10^{-34} joule-seconds, has the much smaller value of 6.6×10^{-68} joule-seconds. What effect would this be expected to have on the scientific knowledge of people living in such a universe, as compared with our own?

3. Gulliver in continuing his astronomical travels proceeds next to discover yet another world called Brobdiplanck, in which Planck's constant has the enormous value of 6.6×10^6 joule-seconds. Try to predict what life would be like for its inhabitants.

16   The Heisenberg uncertainty principle

All Faith is false, all Faith is true:
*   Truth is the shattered mirror strown*
In myriad bits; while each believes
*   his little bit the whole to own.*

RICHARD FRANCIS BURTON
The Kasidah of Haji Abdu, El-Yazdi, VI, 1.

The wish to know things seems to be a characteristic human quality; it can be identified at a fairly early stage in a child's development. But the child soon learns also that the available information is carefully circumscribed. And like other cravings or appetites, this one too is susceptible to inhibition. If knowledge or the process of its acquisition brings pain or discomfort, one learns to suppress the craving. Often such repression is imposed by the existing social structure, which regardless of its political complexion quickly discovers that smooth functioning is impaired when everyone asks too many questions; it therefore tends to reward the conformist and punish the questioner. Thus Socrates paid with his life, and in imperial Russia it was a crime to teach peasants to read. In more sophisticated societies this limiting of human knowledge takes the relatively benign form of the cult of the expert. If everyone's thinking is compartmentalized it can be more easily directed, particularly when what one thinks affects his livelihood. Hence most governments seek to restrict the flow of critical information to those properly conditioned to receive it, and some even parcel it out with a dropper as if it were a dangerous drug. When unpleasant questions are raised the answer may take the form, "Sorry, you simply do not have all the facts; you must trust the judgment of those who do."

By the time a student has finished what we call a secondary school education he has already learned to restrict his curiosity and even derives from such compartmentalization a measure of comfort, as evidenced by the plaintive question, "Why do I have to know *that?*"

Of course specialization is a pragmatic necessity; furthermore, as long as an existing system has to cope with basic problems of survival it inevitably seeks stability and "order," compensating the population for loss of freedom by providing a measure of security. The affluent societies, however, are finding it increasingly difficult to maintain this arrangement, and are seen instead to be groping for something consistent with the fact that more and more of the "peasants" are learning to read.

Fortunately (or unfortunately, depending on one's point of view) the craving to know things has never been entirely suppressed; some very effective measures for accomplishing this were applied during the Middle Ages, and still we managed to come through with only a loss of time.

It is inevitable that in the search for knowledge one should eventually get around to asking whether there is indeed something to know. And if there is, can we know it? And if we can, are there also other things which we cannot know? It was discovered that such questions are more difficult to answer than all the others. One is easily led into circular arguments, and there is actually no reply to the solipsistic philosophy; thus

> "We are such stuff
> As dreams are made on, and our little life
> Is rounded with a sleep."

Or

> "Life's but a walking shadow, a poor player
> That struts and frets his hour upon the stage
> And then is heard no more: it is a tale
> Told by an idiot, full of sound and fury,
> Signifying nothing."

To this Descartes replied, "God cannot have deceived us." *Cogito, ergo sum.* I think, therefore I am. This philosophical approach has sometimes been called metaphysical realism. Its dogmatic flavor and bootstrap-lifting logic were rejected by empiricists like Hume, Locke, and Berkeley. Hume claimed that the only knowledge is that based on direct experience, and raised serious questions about the meaning of words like "existence." Kant subsequently combined these points of view, concluding that there are two types of knowledge, one founded on experience and another which is *a priori*.

As far as the physicists of the classical period were concerned, science had as its purpose the study of the "objective" world, and this was entirely susceptible to discovery. It could be that there were things outside this world, but if there were they were not the province of science. Classical physics was based on materialism, and simply accepted that "things exist." When a tree falls in the forest it makes a noise whether or not there is anyone to hear it. The observer may take note of the tree; but the tree is there in any case.

The development of a scientific method was characterized by three stages. In the first man learned to observe the world around him; thus the ancient science of astronomy was born. Then much later came the contrived experiment, and one was no longer constrained to study the wares nature herself had elected to display; now it became possible to ask questions of one's own choosing and to *extract* the answers from an indifferent universe. And finally there came theory and the tools of mathematical analysis; in this most sophisticated stage order was created out of chaos and laws of nature were projected, including the ability to predict what had never before been observed.

But in all this there was one important understanding; the observer to the event must remain a silent witness. It was the "real" state of affairs which was the target; the purpose was to discover how things acted when they were not being observed, and an objective description of the world was one which necessarily excluded the observer. Now as long as science consisted only of observation and cataloguing this posed no great problem. But with the introduction of the experimental method it became necessary to devise techniques whereby one could tread gently and observe without creating too much of a disturbance. After all, one is not interested primarily in the measuring apparatus, but in the phenomenon as it was before the introduction of the apparatus. And herein lies the basic contradiction—or at least severe limitation. For how does one observe the way things act when they are not being observed?

Suppose, for example, that a woman is very beautiful in a tenuous sort of way: As long as she remains unseen and unrecorded her beauty is such as would make the gods cry out with joy (if they could only know); but exposed to the light of day it fades instantly, giving way to a great ugliness. Now it is easy enough to check her out when she is in the ugly stage; but how do we verify her beauty? In the operational sense of course the question has no meaning; and after our experience with the orbiting astronaut who wasn't there we look upon such hypotheses with great skepticism. But the parable suggests that at least in principle there may be limits to human knowledge and our facility for objectivization.

Perhaps the reader considers this to be hairsplitting. And so it is; but we shall find it necessary to split a number of hairs before explaining the findings of the last chapter. We therefore consider yet another parable: A small new nation has just been created from the ashes of an old colonial empire, and there are conflicting reports about its democratic processes. You are a reporter who has been assigned the task of determining whether or not there is genuine freedom of speech in this country. The problem is that every previous reporter has been provided with an interpreter and a tour guide, and his interviews with the population have invariably evoked great enthusiasm for the government in power. This is consistent with expressions of support in the news media. There is simply no opposition; perhaps some will eventually develop, but so far the government has managed to obtain a perfect consensus.

It would, however, be interesting to observe this small nation under rather more objective scrutiny; namely, people must not know that they are being observed, and are therefore more likely to reveal their true opinions. In order to achieve such conditions it is necessary that (a) you learn to speak the language without trace of an accent, (b) your appearance is so disguised as to be indistinguishable from that of the natives, and (c) you sneak across the border at night without being discovered. All these requirements may presumably be met. The question is, can you get away with it? Is it possible to observe this nation in its "natural" state, namely, as it is when it is not being observed by outsiders?

There is one detail which has not been mentioned: How small can a small nation be? If upon arrival you discover that there are several million people, the plan has a fair chance of success. On the other hand, if there are only a few thousand inhabitants it may become a bit tricky. With a little luck, however, it may still be possible to hide in the population. What if there are only a hundred people? Everyone is likely to know everyone else; it will be difficult to pass as just another native. Perhaps it can be managed for a short time, but eventual discovery is inevitable. Worse yet, what will you do if there are only *ten* inhabitants in this country? It would be a miracle to "pass" even for a moment in such an environment. And now we come to the worst of all possible circumstances: Suppose this "nation" has only *one* native! Not only is it impossible for a reporter to speak to such a "population" without being discovered, but the very questions he would want answered have lost their original meaning. Words like democracy, consensus, and freedom are no longer relevant. It becomes necessary to ask different questions, to employ new concepts, and to discard old pictures in describing such a situation.

In general the assumption that it is possible to know the "real world" without including oneself in what there is to know is an *approximation,* valid only to the extent that the disturbance introduced by "asking the question" is small compared to the effect being observed. Such approximations are very useful, for they are our only source of knowledge; but one must never lose sight of the fact that the accuracy of the information is necessarily limited.

Until the twentieth century the scientist was in a position analagous to that of a reporter who had entered a *large* country with a *large* population, and was observing it without disturbing too much that which was being observed. As an unobstrusive spectator he saw no evidence of any limit on the ability to acquire knowledge. What he saw was a world in which effects followed causes and events ran like clockwork. Such a *deterministic* picture necessarily implies that all information is available, at least in principle. Thus if one can somehow manage to acquire enough knowledge about the present it should become possible to determine everything about both past and future. Scientists were aware of the practical limitations of such a grandiose scheme, but this was a matter of degree;

in principle the sky was the limit. Within the framework of classical physics, if we want to know whether it will rain tomorrow, we simply proceed to collect enormous volumes of data about the movement of every molecule of air and water on this planet; then we can determine what the weather will be not only tomorrow but for all time to come. Of course we are still in a relatively primitive stage of development. Our collection devices are simplistic, our computers limited in the amount of data they can process, and even our physical theories and mathematical skills in need of further refinement. But, said the classical physicists, at the present rate of progress, if the race lasts long enough, who knows? There should be nothing beyond our reach.

Then came the strange atomic experiments, and the questions became more detailed. What does an individual atom look like? What happens when it is divided into its components? Can these components be further subdivided? Why do particles act like waves, and waves like particles? The classical pictures which had so effectively accounted for the behavior of blocks and pulleys and steam engines were no longer able to provide answers. The balls in a bowling alley of "ordinary" macroscopic proportions behaved "normally"; but "balls" of submicroscopic proportions in a submicroscopic "bowling alley" behaved instead like those in our parable—passing through two holes at the same time, eluding our view, giving contradictory answers. Observations pursued on a submicroscopic level revealed that the laws of nature (as designed by people, of course) had simply stopped working. Like Alice walking through the looking glass the human race had entered a world in which apparently utter nonsense prevailed.

The reader may be tempted to question the accuracy of some of the experimental evidence. I can only assure him that there is no such easy way out; laboratory techniques have been brought to a remarkably high level of precision, and equipment inaccuracies demonstrated to be orders of magnitude smaller than the quantities being measured. No, something far more serious had occurred than a faulty piece of apparatus or a misplaced decimal point.

The human race had come up against a problem resembling that of the reporter in the microscopic country. Our explorations had reached the point where we could not conceal ourselves from what we were observing; and our theories of what was real could therefore no longer exclude the act of observation. There is no known reason why the same effect should not qualitatively also be present on the *macroscopic* level; but quantitatively it had never before needed to be considered.

It must be understood that this new complication is not one imposed on man as a result of his frailty or primitive clumsiness; in fact it matters little whether the "observer" is the earth scientist himself or an extremely sophisticated piece of apparatus—or indeed whether it is merely the brute unleashed force of nature "doing her thing." The very process of interaction between the phenomenon and the means for observing it influences that which was to have been observed. And the net effect of such influence may be shown to be a fundamental *uncertainty* in available knowledge, as well as a basic inability to predict future behavior.

In order to demonstrate this restriction on knowledge in the abstract sense, we now take leave of complex sociological parables, and consider instead a particular physical problem which goes to the very heart of the matter. It is only one problem, but its implications are universal, for it leads us directly into the Heisenberg Uncertainty Principle.

Spectroscopic analysis of the light emitted by excited atoms had in the first decades of this century tended to support Bohr's atomic model of the electron as a particle revolving in circular orbit around the nucleus. But as the experimental techniques improved new lines were discovered in the spectrum of hydrogen which could not be accounted for by Bohr's theory. These new spectral lines (called the "fine structure" of hydrogen) were interpreted by Sommerfeld in 1916 as representing changes in atomic state involving not merely Bohr's original circular orbits, but additional *elliptical* orbits as well. When a planet or comet in an elliptical orbit approaches closer to the sun its speed increases, and as it travels farther away from the sun the speed decreases. The same effect appeared to be taking place in the electrons orbiting the nucleus. Sommerfeld discovered that Bohr's rule for quantization of angular momentum, as expressed in equation (15–3), when extended now also to elliptical orbits, seemed to account for the fine structure in the hydrogen spectrum.

It appeared that people finally knew what an atom looks like. One could actually draw these "allowed" Bohr–Sommerfeld orbits and specify precisely what the velocity of the electron would be at every stage of its trajectory. Furthermore, even though no one had as yet succeeded in building a microscope good enough to enable him to *see* the orbits, their energies could at least be computed and were found to agree with the spectroscopic observations to a large number of decimal places. Atomic physics had apparently reached a level of precision comparable to that of astronomical science. In 1845 the French astronomer Leverrier had predicted the existence of the planet Neptune by observing its perturbing gravitational influence on the motions of the other planets. He was thereby able to tell everyone where Neptune would be at a particular moment, and when the telescopes were pointed to the appropriate part of the sky, there it was.

Of course in the case of atomic physics there was the matter of the inexplicable *discreteness* of the atomic orbits, which permitted only certain states to be occupied, according to equation (15–3); but this would presumably someday be explained. (De Broglie was still a schoolboy in the days of the Bohr–Sommerfeld atom, and no one suspected that a material particle would eventually turn out to be a wave.)

If astronomers can point telescopes at the predicted position of a planet, why should atomic scientists not one day point extremely powerful microscopes at the atom and actually *see* the electrons spinning around the nucleus in their tight little orbits? If such a thing were possible, it could unlock for us the most basic secrets of nature, including those of life and death, for biological processes must begin ultimately on the atomic level. At some stage of its existence a healthy cell becomes a cancerous cell; and this cell consists of complex molecules, which are themselves comprised of atoms, which are in turn composed of electrons and protons and neutrons. How deeply we can probe with our microscopes may well determine the ultimate limit on how much we can ever know about ourselves. And it is the business of physics to explore such fundamental questions.

Since the time of the Bohr–Sommerfeld theory there has been considerable progress in microscopy. The electron microscope, which uses *electron waves* instead of light waves, now does enable us to observe some of the larger complex molecules. Will our descendants one day sit in a theatre watching actual films of electrons circling the nucleus of an atom? It is my sad duty to inform the reader that such a time will never come—neither for our descendants nor for any other race in the universe. We shall shortly see why this is so and thereby discover an ultimate limit on how much can be known about the physical world.

Let us suppose in the distant future the allocation of extremely large sums of money for the purpose of photographing the atom; all the important social problems will have been solved by then, and annual budgets which are today allocated for war, space exploration, medicine, high energy particle accelerators, as well as those yet to be appropriated for solving the problems of poverty, ignorance, and pollution, are all funneled into this one microscope project. In order that the scientists should not feel unduly pressured by the enormity of the responsibility, we can give them a thousand years to complete the project, or ten thousand, or a million—whatever is needed.

Figure 16–1 is a representation of the device they may finally come up with. Light of an appropriate frequency illuminates the electron of a hydrogen atom. Upon rebounding off the electron this light is passed through a lens (or system of lenses) of suitably large magnification and brought to a focus on a screen or photographic plate. If a camera is to record the picture, the shutter speed will of course have to be rather high to avoid "smearing" the action shots, since the time of rotation of a Bohr orbit is typically 10^{-16} second. However, we are not here concerned with technological problems; these should eventually all be overcome.

What worries us is the existence of some rather more fundamental difficulties. The first of these is known in optics as the *diffraction limit*. We have already seen that light passing through *two* apertures can produce interference fringes; but it may likewise be shown that even a *single* aperture, when its width is greater than the wavelength of the light, can produce a characteristic diffraction pattern. (As seen in the Appendix, the two halves of a single slit in effect act like interfering sources.) The reader was deliberately spared this added complication in the discussion of two-hole interference; single-aperture diffraction is in fact avoided in such

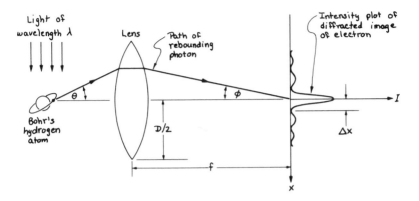

Fig. 16-1 The world's most expensive microscope—cost, 1000 GNP's.

experiments by insuring that the apertures are much smaller than the wavelength. However, now we must face this problem squarely. The microscope lens is itself a large "aperture," since only that light which actually passes through the lens is brought to a focus on the screen or plate; light beyond the edge of the lens is lost, just as if it had been blocked by a barrier.

The light intensity variation in the diffracted image of the electron is plotted at the right of the figure. Since what we wish to photograph is the motion of the electron, we should have liked its image to appear as a sharp clear dot on the photographic plate. But because of the diffraction effect introduced by the lens "aperture," the image of the electron is spread over an intensity pattern of alternating bright and dark fringes. This is represented by the curve of light intensity I versus distance x from the center or "correct" position of the electron image. Most of the photons which succeeded in illuminating the electron fall in the vicinity of the central maximum, represented in the figure by a large peak. To either side of this maximum is a dark region of destructive interference (the first minimum), where the curve dips back down to the vertical axis. The smaller wiggles represent the subsequent maxima and minima.

The size of this diffraction "blur" may be measured roughly by the distance, which we call Δx (delta x), between the center of the pattern and the first minimum. In terms of the geometry of the optics the distance Δx may be shown (see Appendix) to be approximately

$$\Delta x \approx \frac{\lambda f}{D} , \qquad\qquad (16-1)$$

where λ is the wavelength of light, D the diameter of the aperture or lens, and f the focal length. This blurring effect is fundamental and results simply from the basic *wave* nature of the illumination. Of course real lenses normally introduce *additional* aberrations, but we are conjecturing that our scientists will have solved all the practical technological problems.

In order to locate precisely the position of the electron in the photograph, we want this diffraction blur to be as small as possible, so that it will appear most nearly like a point instead of a smear. Therefore, if we wish to minimize the undesirable effect, we must make the denominator D of the above expression as large as we can. (The larger the denominator of a fraction, the smaller the fraction.) This is in fact one reason why the telescope at Mt. Palomar has a mirror (corresponding to our lens) two hundred inches in diameter. Since money is no object in our project, the scientists may possibly come up with an enormous microscope lens, perhaps hundreds of miles in diameter. Also, from equation (16-1), another way to keep Δx small is to make the focal length f in the numerator as *small* as possible (the smaller the numerator, the smaller the fraction), and likewise to employ light of the smallest possible wavelength λ for illumination, namely, ultraviolet light.

The diffraction problem is clearly a nuisance, for it can turn the clear-sighted scientist into a myopic observer who sees only blurred images. But of course we can make the blur as small as we like by making the lens as large as we wish, etc. If this were the only obstacle to human knowledge, I should not have burdened the reader with the details of the diffraction problem. Unfortunately, however, there is a second problem, and this one is similar to that of the reporter in the small country. If the electron orbit is to be revealed as it "really" is, it must be disturbed as little as possible by the act of observation. Now if we illuminate it with light of excessively high energy we are likely to knock the electron out of the atom. We shall therefore use as weak a light as possible. What is the weakest possible light of a given frequency? It is just *one photon*. This single photon will bounce off the electron, and with a little luck make its way through the lens to the photographic plate, landing somewhere in the diffraction pattern. The most probable point of its arrival is of course the central maximum, since this is where most of the photons land when a more intense light is used. In fact, if there were no diffraction effect the single photon would have to land precisely in the middle of the pattern. But now all we can say is that its point of arrival is *probably* within the central maximum; hence the "probable" error in position of the image (the deviation from the exact center of the pattern) is roughly of the order of Δx.

But even a single photon can do a lot of damage to an atom when the photon's energy is large enough to be comparable to that of the electron itself (just as our mythical reporter would have messed up the tiny country he was investigating). The question is: *How much does the photon disturb the electron in the process of illuminating it?* This of course depends on whether it is a head-on collision or only a glancing blow. In our investigation of the Compton effect we were in a position to answer such questions by placing detectors at various angles to catch the photon (thus determining its direction) after the collision. But now it is the lens itself which serves as the detector. The very fact that the photon appears at the screen tells us that it must have passed through the lens and been brought to a focus. But we have no way of knowing *which part* of the lens the photon passed through—whether the center or the periphery. Only by making the lens very small can we restrict the angle θ in Fig. 16–1, and hence minimize the uncertainty in the direction of the one photon which gets through. (We assume all additional photons have been eliminated.) In fact if we had our way we should use a lens whose diameter is practically zero! Then there would be no uncertainty at all. And once we knew the direction of the photon we would thereby know the precise extent of the disturbance to the atomic electron. This would be important in interpreting such photographs and deducing what the electron's orbit would have been if we had *not* disturbed it.

But do we dare use such a tiny lens? We do not. The lens designers would hit the roof, for they know only too well the nature of the diffraction problem, and the fact that a *large* lens is required (as we have seen) to keep the uncertainty Δx in the *position* of the electron from getting out of hand. Reducing the uncertainty in the position of the electron increases the uncertainty in the extent of the disturbance; and reducing the uncertainty in the extent of the disturbance increases the uncertainty in position.

This is the nature of the universal paradox embodied in the *Heisenberg Uncertainty Principle,* which we see here in only one of its manifestations. Let us summarize the situation in quantitative terms. It is shown in the Appendix from geometrical considerations that the uncertainty Δp (delta p) in the x-component (up-down) of the photon's momentum, and hence in the momentum of the electron as a result of its having been photographed is given by

$$\Delta p \approx \frac{h}{\lambda} \frac{D}{2f}. \tag{16-2}$$

This minimal uncertainty in the amount of the disturbance results from the *particle* nature of the illuminating light. Practical considerations might introduce more than one photon and make matters worse, but we are giving the scientists the benefit of all doubt.

To optimize our knowledge of the *momentum* of the electron, and hence our ability to correct for its having been disturbed by the act of observation, we want large λ, large f, and small D, for this will minimize Δp. But to optimize our knowledge of the electron's *position,* from equation (16-1) we want small λ, small f, and large D, since this minimizes Δx. The only remaining possibility is the reduction of Planck's constant h, which would require a redesign of the universe; this is of course no alternative at all.

How are we to establish whether the scientists will have done a good or a bad job after spending a thousand or more GNP's (gross national products) on this undertaking? What is required is that *both* Δx and Δp be as small as possible, since both are required to enable us to interpret the photographic observations. Hence an appropriate figure of merit for the success of this microscope project is how small the *uncertainty product* $\Delta x \, \Delta p$ turns out to be. It is easily seen that regardless of what decisions they eventually make with respect to lens diameter, wavelength, and focal distance, when it is all over, the very best they can hope to achieve is

$$\Delta x \, \Delta p \approx \frac{\lambda f}{D} \frac{h}{\lambda} \frac{D}{2f} = \frac{h}{2}, \tag{16-3}$$

since D, f, and λ may be canceled in this product.

The calculations in the Appendix which produced equations (16-1) and (16-2), and hence (16-3) were actually rough approximations. A more careful statistical analysis (which considers the probability of the photon's having passed through various portions of the lens, etc.) results in the precise statement of the Heisenberg Uncertainty Principle, namely,

$$\Delta x \, \Delta p \geqslant \frac{h}{4\pi}. \tag{16-4}$$

What this says is that the expected uncertainty Δx in a particle's position, multiplied by the uncertainty Δp in its momentum, must be at *least* as great as Planck's constant h, divided by 4π. With the best of all possible microscopes, designed by the best of all possible scientists in the best of all possible worlds, the product of the two uncertainties is $h/4\pi$; practical limitations which we have not considered here, like the problem of casting a perfect lens, for example, may of course increase the uncertainty product. But $h/4\pi$ is the absolute minimum which can ever be achieved.

We have here considered a particular example, but it turns out that a similar restriction appears in *all* such problems. Whether one is trying to design the best of microscopes or undertaking any other experiment which requires measuring simultaneously the position and velocity of an object, the minimum uncertainty product always makes its appearance to limit the available knowledge. It does not matter whether a photographic record is produced or an animate observer views the event directly. Nor does it matter whether in the design one uses visible light or electron beams or radar signals, or any other device, wave or particle, to explore the unknown phenomenon. Always the same limit appears. And the key to this riddle of uncertainty is the number known to earth scientists as Planck's constant—a number which he first encountered when he had to quantize an energy source to explain blackbody radiation, which Einstein then discovered to be the basis of the particle behavior of light in the photoelectric effect, and which now we too find to be determining the least possible quantum of accuracy in the uncertainty product when we try to design a perfect microscope. Obviously it is a very important number indeed; for it sets the threshold of available knowledge. We can know exactly where a particle is, but then we cannot know at all which way (and how fast) it is going. Or we can know exactly which way it is going, but then we cannot know at all where it is. Or we can know a little of both where it is and where it is going—but then only so much as Planck's constant permits. And the restriction is a universal one. Analysis of the light from distant stars and galaxies reveals precisely the same spectroscopic properties as are found in sunlight. One would thus expect that Planck's constant has been discovered in one form or another on every civilized planet in the universe (as may be seen in Fig. 16–2).

Fig. 16–2 *On a distant planet.*

Let us now return to the problem of photographing the atomic orbits. How does the uncertainty principle affect the ability to "see" an atom? Our scientists of the future will obviously have to design their photographic system to resolve (see distinctly) distances which are somewhat smaller than the diameter of the Bohr atom. This is the minimum required uncertainty Δx in the electron's position. (Otherwise the atom will be effectively one "big" blur.) But equation (16–4) tells us that if, for example, we require $\Delta x =$ (one-fifth of an orbital diameter), then Δp, the expected disturbance or uncertainty in momentum introduced by observing the electron's position to this degree of accuracy, will have to be at least

$$\Delta p \geqslant \frac{\text{Planck's constant}}{4\pi \text{ (one-fifth of an orbital diameter)}} . \tag{16–5}$$

This result is obtained by dividing both sides of equation (16–4) by Δx. When actual dimensions are introduced (see calculations in the Appendix), the momentum uncertainty Δp which is consistent with "seeing" the atom turns out to be essentially of the same order of magnitude as the original momentum of the electron before we tried to photograph it! In other words, the effect of the least disturbing usable photon is great enough to knock the atomic electron right out of its shell; and because of the momentum uncertainty we cannot even know whether what we are observing is an electron which is still in orbit or has already begun traveling through space. Translated into practical terms, what this means is that the only way to "see" an atom is to destroy it.

It follows that, although one can in principle design a microscope to photograph an atom, it should be expected to be the last such photograph ever taken of this particular atom; and when the electron's position is determined with enough precision to be interesting, there is absolutely no way to obtain a useful estimate of the resulting change in its *velocity*. On the other hand, if the design parameters (lens diameter, focal length, and wavelength) are so chosen as to leave the electron relatively undisturbed, then its *position* in the atom will necessarily be blurred over a region comparable to the orbital diameter.

There is no way to "see" the atom in operation; it does not matter whether the "observer" is a human being or a piece of equipment or just the stars twinkling in the sky. Nor does it matter whether the observation was originally arranged by an intelligent and purposeful creature or is simply a whim of nature. Information about position always interferes with information about velocity. You can know one precisely, or you can know the other precisely, but you can never know both precisely together. There will be no movies of the atom, not today and not in a million years.

This joint uncertainty in position and velocity of a particle means that it is not possible to predict exactly what the future will be; for precise knowledge of future position demands precise knowledge of *both* present position and present velocity, and these are not simultaneously available. The very act of acquiring complete information about events in the present invariably introduces an uncertainty about those in the future; in the case of the atom precise knowledge of present position *destroys* its future.

We can of course see any one particular aspect of the physical world with arbitrarily great precision and clarity; but then certain other aspects become muddied and uncertain. Our quantum mechanical reality is indeed a "shattered mirror strown in myriad bits." And neither we nor anyone else can ever put the pieces together again.

The deterministic picture of the universe has failed to survive our investigations; for determinism means that the seeds of the future are contained within the present, and since the present can never be fully encompassed, in the true operational sense no precise causal connection can ever be made. Of course this is not to say that knowledge of the future is completely lost; it is only somewhat blurred. We cannot predict precisely which event will occur tomorrow, but, like the weather man, we can predict the *probability* of an event. Moreover, this blurring of total knowledge affects not only future events but even those in the present.

Figure 16–1 presents a detailed plot of the probabilities of various "events." The light intensity curve, which shows maxima where photons passing through the apparatus are most likely to land and minima where virtually none of them will land, may be interpreted as a probability curve. The intensity of light, or number of photons arriving at a given point in a given time, is necessarily proportional to the *probability* that a *single photon* will arrive at that point on the screen; the high probability spots are the ones which receive the most photons, and the low probability spots the ones with the fewest photons. Thus when we illuminate the atomic electron with a *single* photon there is no way of knowing *a priori* where the image will appear with respect to the "true" position; but it is possible to know precisely the *probability* that it will deviate from this position by any particular amount.

Poet Why can I not arrange to have the photon glance off the electron at just the right angle to permit it to arrive in precisely the center of the intensity pattern? This would be like calling a shot in billiards; if I am a good enough player I can make the ball land in the side pocket. Your uncertainty about the electron was apparently due to an original uncertainty about the exact position of the incoming photon. But if I really call my shots I can predict how much the electron will be "disturbed" by the photon, because I know whether it will be a direct hit or a glancing blow. In this way I should be able to reconstruct both the position and velocity of the electron before the collision, and thereby predict what they will be in the future.

Scientist You are still drawing pictures of a deterministic world. There are two things wrong with this plan. In the first place we do not know in advance exactly where the electron is in its orbit; remember, this is one of the things we set out to determine. So we cannot predict head-on collisions or glancing blows. And secondly there is no way of *preparing* the photon so as to assure that it arrives at precisely the required position (even if we knew it) with precisely the required momentum. Choosing the wavelength of the light by passing it through a filter corresponds to specifying the momentum of the photon, in accordance with equation (14–10). Just as we cannot simultaneously determine exactly the position and momentum of an electron, so the same fundamental obstacles invariably prevent us from *preparing* (or *knowing*) simultaneously the position and momentum of the photon. Accurate billiards is all very well for macroscopic players using macroscopic balls in a macroscopic world with a small value of Planck's constant. But in this game our player is too blind or too nearsighted to see the ball clearly; and there is nothing anyone can do to restore his vision. Whenever we try to do so we invariably wipe out the game.

This probabilistic version of reality (and it is by now perhaps evident to the reader that this is where our investigation has been taking us) is the direct result of one's inability to specify simultaneously and with infinite precision such things as position and velocity of a particle in a given direction. (Strictly speaking we should say "momentum" rather than "velocity," since then we include massless particles like photons.) There are also other pairs of "observables" which obstruct each other when we try to measure them simultaneously; like position and momentum, so also information about the exact amount of *energy* produced in a phenomenon (such as the radiative decay of an excited atom) and the precise *time* at which it occurs is likewise limited by the Heisenberg Uncertainty Principle. Such pairs of interfering observables are called in physics *conjugate variables*. The reader is warned not to make reckless generalizations about the uncertainty principle. There is much which *can* in principle be known exactly; only certain specified properties of a system obscure each other when observed at the same time, and these are the conjugate variables.

Thus sweeping statements like "Nothing is certain" are quite meaningless; nature has elected to confine us with its own particular restrictions. If we may return once more to allegory, let us suppose that an airplane is flying over what we believe to be the exact center of Chicago's Midway Airport at what we like to think is a precisely specified velocity, and we want to predict therefore exactly when it will arrive in Los Angeles. Unfortunately the knowledge that it is over Chicago has been obtained as the result of illumination by sunlight, or radar beacon, or perhaps some other device not yet invented; and such illuminating photons (or other carriers of knowledge) have in principle necessarily disturbed the velocity of the airplane, taking it off course by an uncertain amount. We can know that it is precisely over the center of Chicago; or we can know that if it were over the center of Chicago it would be headed in precisely the right direction at 700 miles per hour; but we cannot know *both* of these things. Hence we cannot predict with absolute certainty that the plane will ever arrive in Los Angeles. Even if the pilot tries to bypass the uncertainty principle by signaling us himself, there will be a joint uncertainty in the *energy* content of his signal and the *time* he sends it which will prevent us from knowing all the relevant information.

Fortunately for the airlines, Planck's constant is small enough and the airplane massive enough to permit such factors to be disregarded in publishing predicted arrivals; else the flight schedules would have to be replaced with probability curves. Because the effect is so negligible for objects like airplanes or persons or billiard balls, we have managed to convince ourselves that we live in a deterministic and "objective" world. What we call objective reality is a macroscopic universe peopled by macroscopic objects; and what we see as deterministic behavior is evidently only the most *probable* behavior of large masses or large ensembles of particles. But we must never forget that all truth is not contained within such descriptions.

questions for the reader

1. The diameter of the earth's orbit around the sun is some 100 million kilometers, its orbital velocity is 30 kilometers per second, and its mass 6×10^{24} kilograms. If the entire solar system were suddenly transferred to the mythical kingdom of Brobdi-planck, where $h = 6.6 \times 10^6$ joule-seconds, how would this affect the chances for an astronaut to find his way home from a trip to Jupiter? Compare the relevant planetary parameters with those of the Bohr hydrogen atom (orbital diameter = 10^{-8} centimeters, electron angular momentum $mvr = h/2\pi$), thus determining whether the space program would have to be terminated as a result of this revision of Planck's constant.

2. a) A particle is ideally "prepared" by being given a *precise* value of momentum; i.e., the momentum uncertainty $\Delta p = 0$. How accurately is it possible to know its position?

   b) A similar particle is localized at a precise point in space; namely, we have ourselves placed it there, so we know exactly where it it at that instant. What can be said about the momentum of this particle?

17 The world of quantum mechanics

Commend me to them;
And tell them that, to ease them of their griefs,
Their fears of hostile strokes, their aches, losses,
Their pangs of love, with other incident throes
That nature's fragile vessel doth sustain
In life's uncertain voyage, I will some kindness
  do them:
I'll teach them to prevent wild Alcibiades' wrath.

WILLIAM SHAKESPEARE
Timon of Athens, Act V, Sc. 1

Now that we have been properly jarred out of the state of mind of one who knows not and knows not that he knows not, we have earned the right to know something of what we can know. The uncertainty principle and wave-particle duality have doubtless sown a measure of confusion; it is time to restore some of the fences.

Actually, all is not lost. It is true that waves are particles and particles are waves, that position and momentum cannot simultaneously be precisely defined in the operational sense, and that there is no way to predict the future with certainty. All this would be a traumatic discovery for a young child who needs a simple model of the universe. But the human race has presumably at least entered its adolescence, even if there are still some big hurdles to be negotiated before attaining majority. We are a long way from the predestination of the ancient Greeks who were playthings of the gods. Uncertainty has been introduced into our lives. And in discovering some of our limitations we have also gained a measure of strength, including incidentally the power to destroy ourselves. We like to believe that we are fast attaining means for curbing the "wild wrath" of nature, even if we have yet to find a way to curb our own.

On the other hand one should not underestimate the bewilderment introduced by the wave behavior of particles. It would be easy enough to dismiss the interference experiments if they were restricted to electrons or any other special class of particles, just as the particle behavior of light waves was once thought to be a peculiar property of electromagnetic phenomena. After all, electrons are charged particles which radiate electromagnetic waves when subjected to accelerated motion; thus it might appear conceivable even in the classical sense for them to exhibit wave properties under special circumstances. But the phenomena described in the particle interference experiments are not restricted to electrons or even to charged particles in general; in fact they are apparently not restricted at all. The interference fringes have been observed with all kinds of particles—like neutrons, which carry no electric charge and interact only upon very close proximity. Moreover, the width of the fringes in a particular experiment specifies the wavelength of the particles in accordance with equation (13–1), which was originally derived for light waves. The size of the fringes therefore depends through the de Broglie relationship only on the momentum (mass times velocity) and on *no other property* of the particles used in the experiment. Thus what we are seeing appears to be a universal manifestation of an effect which is never observed in the case of "ordinary" macroscopic objects only because their masses are too large to produce visible fringes. It is this general property of physical reality which is therefore a principal object of our investigations.

The early successes of the Bohr and Sommerfeld models of the atom were destined to be short lived. Sommerfeld's calculations applied special relativity to obtain energy corrections for the elliptical orbits, and his agreement with the spectroscopic data for the hydrogen fine structure was truly astounding in view of the fact that the atomic model turned out to be quite wrong. When more complex atoms were examined, and electric and magnetic fields introduced, the Sommerfeld approach was unable to explain the experimental data. And no amount of patching could make it work. We know today that the Bohr–Sommerfeld model was the last stronghold—the best that could be done to explain atomic structure by means of traditional pictures. By the middle nineteen twenties it had become abundantly clear to a number of people that atomic reality was not susceptible to classical models, and that a radically different approach was necessary.

It was the de Broglie hypothesis which suggested the new description of nature; and an important clue is the failure *in principle* of any attempt to "see" the electron traveling around the nucleus. If something cannot be seen even in principle, how can one expect to describe it in terms of pictures which *can* be seen?

Thus the world of quantum mechanics is not one which an artist can put to canvas (at least not in the representational sense). Our model of this world cannot *look* like anything at all. It should merely have the ability to generate pictures when we ask questions which can be answered in pictures; in other words, it must predict correctly the results of all possible experiments. But we cannot expect to put the pictures together into one big picture, for the uncertainty principle precludes such simultaneous viewing. Like the fragments of the shattered mirror, "each believes his little bit the whole to own." Let us examine the shape of this quantum mechanical world.

First we resolve the paradox of wave-particle duality. It would appear to be a contradiction that the same object should be both localized in space and also *not* localized in space. Like all apparent paradoxes it arises from a limited view of things. When our senses interact with the world, our memory banks store the impressions, and the mind then calls upon these stores to "explain" subsequent impressions. The bigger the store of old memories the more convinced one is that there is nothing new under the sun. When something comes along which cannot be interpreted in terms of the old pictures (and there must be many such phenomena in the universe), the mind nevertheless tries to make it fit. If one day the eyes see a unicorn, the mind says, "What's a unicorn?" and the memory banks say, "It's sort of like a horse." Whereupon the eyes say, "But look at the horn coming out of its head," and the memory banks say, "Well, now, I guess it's really more like a rhinoceros." The mind then concludes, "Sometimes it's a horse and sometimes it's a rhinoceros; really it's a paradox."

The particle and wave aspects of a phenomenon need not be mutually exclusive. In the language of quantum mechanics they are merely *complementary* pictures of a reality which does not always lend itself to pictorial representation. Such complementary descriptions should be regarded as equally valid, since they are simply alternate aspects of the same reality. The two are of course directly related (as in de Broglie's expression $\lambda = h/p$) by Planck's constant. In terms of characteristic dimensions of our bodies and of the events we observe in the everyday world, h is an extremely small quantity; in fact it is so small that it could easily be taken to be zero. Hence early man (before this century) lived in what physicists would call a classical world. In such a world the wavelengths associated with macroscopic objects (like bowling balls and railroad trains) are so small as to be entirely beyond the range of observation. Thus in the world of classical physics a particle is a particle and a wave is a wave, and there are no paradoxes.

But with the photoelectric effect and the diffraction of particles we enter the twilight zone. Photoelectric phenomena make their appearance under threshold conditions which force light to reveal its particle aspect, and submicroscopic particles impinging on a submicroscopic atomic lattice spacing produce visible wave fringes. Hence we must conclude that the basic phenomena are in general not just waves or just particles, but rather wave-particles.

We have seen that the waves in a sense "tell" the particles where to appear, and that de Broglie therefore called them "pilot waves." Since one object of physics is to predict the results of experiment, knowing in the case of ordinary particles what the *waves* are doing not only tests the theory but also suggests experiments never before performed and reveals a world not yet seen. The Schrödinger equation of quantum mechanics is a *wave equation* which prescribes the *amplitude* of the wave throughout space and time. Schrödinger was conversant with problems of acoustics, and he recognized from the similarity of mathematical form that the behavior of "bound" particles like electrons in an atom resembles that of *standing waves* in an enclosure (like a speaker cabinet), whereas the behavior of "free" or "unbound" particles is associated with *traveling waves,* such as radio signals propagating through space. From analysis of the detailed experiments he deduced or "derived" an equation which predicts how the "matter waves" behave when the particles are subjected to arbitrary forces or barriers. What is truly amazing is that its predictions have always been confirmed in countless experiments of atomic and nuclear physics over the entire domain of applicability.

We have still said nothing about what it is that is waving when a beam of particles approaches a barrier, or electrons circulate around the nucleus of an atom. It is one thing to say that particles are waves and another to understand what is meant by the statement.

The solution of the Schrödinger equation is called a *wave function* of position and time. What do we mean by a *function?* It is a mathematical representation of something like a contour map, which shows all the hills and valleys in the terrain. A view of such a map tells us for every point on the ground what the height is above sea level. We may choose to build such a terrain map in the form of a relief structure which literally models the heights (amplitudes) of the hills and valleys (waves in the ground surface). Alternately a mathematician may seek to convey similar information in the form of an equation which describes this surface. The solution of the equation is called a *function* (height above sea level) of *position* (latitude and longitude, or *x* and *y*).

Suppose now that while we are studying our contour map a volcano erupts somewhere, accompanied by earthquakes and a realignment of land surfaces; some valleys become hills and hills become valleys. If the designer of the relief map was sufficiently foresighted, he might have anticipated this cataclysm by constructing the relief map of plastic material which can be remolded to follow changes in the land surface. And if he had been really clever, he might even have mounted under the relief map a system of cams and wheels driven by a computer, which in turn receives data from geodetic survey teams on the scene, thus keeping the relief map up-to-date. It would be a very lively sort of map, but rather expensive to maintain. A much cheaper way to convey the same information would be a representation of the scene by a mathematical function not only of *position* but of *time* as well. Such a function defines the amplitude (height) of the land waves (hills and valleys) at every position on the ground at all times, past, present, and future. Of course the mathematician or physicist who derives the equation which defines such a function would have to know completely the geology of the land—a great deal more information than is presently available about a complex system like the earth. But in principle, if he lived in a deterministic universe, he could collect such information and predict mathematically the precise shape of the earth for all time to come.

The Schrödinger equation does this very thing for the "pilot wave" amplitudes which determine the behavior of a system of particles. If all the forces and barriers encountered by a particle or system of particles are known in the classical sense (as we knew the locations of the slits or holes in the bowling ball parable), then one is immediately able to write the appropriate Schrödinger equation for describing the behavior of the waves associated with this set of particles; the waves then predict what the particles are likely to do when sent against such a barrier or subjected to such forces. Just as the contour map revealed the height of the ground above sea level, so the Schrödinger equation gives the *wave function*, or *wave amplitude* of the pilot waves at all points in space and at all times.

However, we have so far studiously avoided answering the original question: What is the *physical* significance of the wave function? What corresponds to the *height* of water waves and hills and valleys on a contour map, the *density* of molecules in an air pressure wave, the *strength* of the electromagnetic field of force which moves the hammers in the "people wave"? What is waving? It is the same question we once asked about light waves. But it was one of the last questions to be answered, for it was in the reply to this question that the human race truly entered the strange world of quantum mechanics. It was this last step which Einstein refused to take, even though there was in fact nowhere else to go.

We have already seen that fringes in the diffraction pattern of light waves are produced by interfering wave amplitudes, and that the light intensity (which can be shown to be proportional to the *square* of the electromagnetic wave amplitude) is an expression of the *probability* of arrival of a single photon at a particular point in the pattern. If interfering electromagnetic wave amplitudes determine the probability of a photon's arrival, why should the pilot wave amplitudes not play the same role for a "material" particle?

It was therefore proposed by Max Born that the wave function, the thing which is waving, which penetrates the slits in a barrier, permeates the space around the nucleus of an atom, and describes the behavior of any system of particles, is a *probability wave*! When this wave function is determined for a system of particles by solving the Schrödinger equation, then the *probability* of finding a particle at a particular position and a particular time, past, present, or future, is given by the square of the magnitude of this wave amplitude. Whereas our contour map was a function whose amplitude gave the *height* of the ground above sea level, the Schrödinger wave function (or rather its square magnitude) gives the *probability* of finding a particle. It was the discovery that light is as much particles as it is waves, combined with the observed wave behavior of "ordinary" particles, which led to the principle of wave-particle duality and the resulting probabilistic interpretation.

This attribution of physical properties to a mathematical concept like probability, and conversely, the suggestion that the ultimate explanation of physical events takes a form as abstract as that of a probability wave, was not something to be accepted lightly. Experimental tests of the new theory were necessary. One of the first problems of interest was that of the electron in the hydrogen atom. When the Schrödinger equation is solved for the hydrogen atom, the resulting wave function determines the probability that the electron will be at various distances and directions with respect to the nucleus. The nucleus is thus enveloped in a *probability cloud*, (the probability of finding the electron), just as the earth is surrounded by the atmosphere. In those places where the atmosphere is denser one is more likely to find an air molecule; similarly where the probability cloud surrounding the atomic nucleus is denser one is more likely to find the electron. One can never say exactly where in the atom the electron is located at any instant—not as long as the atom remains intact. All one can specify is the *probability* that it will be in various places.

Experiments employing both spectroscopic techniques and scattering processes, in which beams of charged particles are scattered by atoms, have invariably corroborated with great accuracy this model of an electron charge cloud, not only for the hydrogen atom but for more complex atoms as well. It is interesting, incidentally, that this electron cloud actually extends all the way from the nucleus out to infinity, although it is most dense (in the ground state of hydrogen, for example) at the very radial distance which Bohr deduced for his first allowed orbit. Hence although there is a finite probability of finding the electron at various distances from the nucleus (including even a very small chance that it will be many miles away), the *most probable* distance is half an Angstrom $(0.5 \times 10^{-8}$ cm), the radius of the first Bohr orbit. Thus it is clearly no coincidence that the Bohr and Sommerfeld models appeared to work so well in the early experiments—just as it is no coincidence that Newtonian mechanics was so successful before the Michelson–Morley experiment. This is, as we have seen, a requirement of every new physical theory: It must enfold within it the predictions of the old theory, explaining why the latter succeeded so well within the range of its experience.

And so a significant break was made between the new atomic physics and the old science of astronomy. Unlike the earth traveling around the sun, the electron does not have a precisely defined orbit; in fact the very notion of such an orbit violates the uncertainty principle and can have no physical basis. Reality is in this case a probability cloud, a smear of charge distributed about the nucleus, predicted by the Schrödinger equation and confirmed with great accuracy in all the experiments.

The wave function or probability amplitude also explains why a beam of particles behaves differently depending on whether there is one open aperture or two, and why a single particle is able to interfere with itself when the waves pass through two apertures. As long as we do not disturb the system by trying to learn precisely where a particle is at a given moment, the probability waves proceed through the apparatus, penetrating all available openings in accordance with the usual behavior of waves. Since this is as true for a high-density beam as it is when only one particle passes through at a time, interference can occur regardless of the number of particles. When we eventually *observe* the particle at the detector (it produces a flash on a scintillator or an image on a photographic plate), it appears or fails to appear in accordance with the rules of probability determined by superposition of those probability waves which have passed through the available apertures. However, if we place a detector in the neighborhood of one of the apertures, or otherwise try to sense the presence of a particle prematurely (before it reaches the screen or plate), we necessarily disturb the probability waves, changing their phase by an unpredictable amount and hence destroying the *coherence* of the beam; since only coherent waves are capable of systematic interference, the very act of determining which hole a particle goes through causes the interference effect to vanish. The same thing happens with light waves; one cannot "see" or otherwise sense electromagnetic waves without inserting a device which interacts with and hence modifies them in some way, destroying the coherence of the beam. By the same token, an attempt to discover the location of the atomic electron invariably disturbs the standing waves of the probability cloud, destroying the atom.

Classically one should be able to employ a sensing device of such a delicate nature that it will barely disturb the state of the system. And this is indeed possible in the case of macroscopic phenomena for which the energy content of information about the event may be made negligibly small compared to that of the event itself. Hence for such systems it is possible to conceive a reality in which the interaction between the event and the apparatus of observation is negligible; this is what is meant by "objective" reality, one which "exists" without being observed. It may reasonably be used to describe the familiar world of macroscopic phenomena. But when we attain a sufficiently detailed microscopic view of nature, then in accordance with the Heisenberg Uncertainty Principle it becomes impossible to obtain information without introducing a disturbance which necessarily changes that which is being observed. If we try to use the approach of classical physics we find ourselves in a predicament analogous to that of the reporter trying to extract information about democratic processes in a one-man country; and, like the woman who insists she becomes beautiful when the lights go out, the question has no meaning.

One may be tempted to think of the electron probability cloud as a "track" left in the atom by a fast-moving particle in a constantly changing orbit. Unfortunately there is a serious objection even to this interpretation. A charged particle moving in any kind of orbit invariably radiates electromagnetic waves as a result of its acceleration (like the oscillating charge in a transmitting antenna). But no such waves are ever observed as long as the atom remains in one of its relatively stable states (called *stationary states*). It is in fact fortunate that this is so, for if the electron radiated away its energy it would have to spiral inward (like a satellite losing energy because of atmospheric friction); the resultant collapse of the atom would leave a world without chemical elements and certainly without people.

Only when there is a *change* in the state of the atom, as when it decays from an excited energy state to a lower state, is there emission of electromagnetic radiation; and the energy of the radiated photon in this case corresponds to the energy lost by the atom. Since in the stationary state there is no radiation, it must be that the electron in the atom is no particle—not until someone (or something) tries to "see" it, whereupon it is indeed revealed to be a particle; but then it is no longer an atomic electron.

The quantum mechanical formalism accounts for such seeming paradoxes, for it incorporates the act of observation in the analysis of the event. But since the available information is never complete, predictions about the results of experiments, like the state of the weather, are expressed in terms of probabilities. The theory is very accurate when it supplies such information, although its manner of providing it is rather paradoxical. In spite of the fact that the old picture of reality is no longer with us, having been replaced by probability waves, both answers and questions continue to be couched in the familiar classical language. This is in part historic necessity, since the language of pictures is the one we know well, but it is also a result of the fact that everything we actually *do* or *observe* is basically macroscopic, even when it is the manifestation of a submicroscopic phenomenon. Thus one formulates a question in quantum mechanics as follows: "Given the following particles with the relevant masses, charges, and other known properties; given also the interactions known to exist among the particles, or between them and the apparatus: What are all the various possible stationary states of such a system? In other words, what probability wave patterns can exist? If the system is initially in one of these states, how long can I expect it to remain in this state before making a transition to another state? When it finally makes a transition, which of the other states is it possible for it to enter, and what are the respective probabilities that it will do so? Suppose I were to determine various observable properties of this system using the most accurate techniques possible. (The 'observables' in question could be the position of a particle, or its velocity, or rate of rotation, etc.) Which of these observable properties can be determined without disturbing the state of the system, and what will be the exact numerical value of the result in each case? For those observables which cannot be determined without disturbing the state of the system, what are all the possible results obtained when they *are* measured, and what is the probability of obtaining each such result?"

Not only can questions of this sort be answered in principle, but whenever the experiments have been carried out they have always confirmed the predictions of the theory. This view of physical reality is of course strictly probabilistic. Because of the uncertainty principle one can never predict *which* of a set of identical excited atoms is going to decay first, but one can say *how many* will decay in a given time, or equivalently, specify the *probability* that any one of them will decay within a given period. Such a restricted description is of course not in keeping with our intuitive belief, fostered by years of macroscopic living, that there must be more to existence than the roll of the dice, the spin of the roulette wheel. "Surely it is only our limited knowledge," we say, "which forces us to describe nature in this way. Some day we shall know what things are really like." And when the teacher has his back turned we try to draw secret pictures of a "real" reality. But the traditional pictures, like the simplistic visions of a schoolboy, become fraught with contradictions under close scrutiny. If the atomic electron is really a wave, why should a sharply resolved "photograph" reveal it to be a particle? And if it is a particle in orbit, why does it not radiate electromagnetic waves? If on the other hand it is a wave when bound to the atomic nucleus, but becomes a particle when we free it by the act of observation, why then does it interfere with itself like a wave upon passing through holes in a barrier, yet flash like a particle when it strikes a scintillator?

And so we try unsuccessfully to fit macroscopic descriptions to submicroscopic events. Yet when the classical representations are transformed by the postulates of quantum mechanics to obtain the probability wave amplitude, a means is provided for predicting exactly the *probability* that a particular picture will be revealed as the result of an observation. This wave function, despite its purely mathematical character, is in a sense more substantial than the pictures; for it is precisely determinable, whereas the pictures appear to be only the results of a game of chance. For macroscopic events of the sort which comprised our experience before we began to look into the atom, the statistical probabilities are so overwhelming as to become virtual certainties; the pictures associated with these almost-certainties are what constitutes our familiar deterministic world. But on the detailed submicroscopic level the mathematical model is the best representation of reality we have, for it successfully predicts pictures obtained as results of all possible experiments; yet it is not itself a picture.

Poet It appears that you are using your own physical theories—in other words, your own state of knowledge, to prove that there is an absolute limit on *all* knowledge. And so you conclude there is nothing but the probability waves. Isn't this a circular argument—in fact rather presumptuous? What happens when a future scientist comes up with a new theory?

Scientist This business of trying to outguess the scientists of the future is as productive as perpetual questions about one's sanity. It is certainly possible that our reasoning is false; or lunatics are right and we are all mad; or we are only figments of a dream; or perhaps our senses were designed to create a deliberately falsified view of reality. These may be interesting questions to contemplate briefly; but I think it is a mistake to become impaled on such questions. One seems to see a big beautiful universe out there, and it is much more interesting to go exploring than to sit contemplating one's navel and asking, "How do I know I really exist?"

P That is of course your judgment. It may turn out that we have more to learn from looking into ourselves than in traveling to the stars. You air of confidence stems from the fact that you have found a formula which appears to work for the time being. Don't forget that you are overriding not only the questions of a poet but the objections of Einstein as well—and he is your own man.

S It is true that Einstein was never able to accept the failure of the mechanistic view. He spent his life looking for an alternative —a new theory which would somehow recover the deterministic universe. But he died without finding it. Perhaps someone else will find it; perhaps not. In the meantime there is a working model which answers questions in complete accord with the results of all observations and experiments ever carried out, which has been dramatically successful in its predictions, and which has provided the basis for a whole new technology. It would be a mistake to dismiss it by saying, "Maybe someday there will be a better model." It is like refusing to ride in an airplane because one day there may be a faster mode of transportation—like transmigration of souls, for instance.

P   Some of us may just not care to take the ride, since we find we are not so particularly enamored of either the brave new technology or the beautiful universe you assure us is being discovered "out there."

The probability wave amplitudes of quantum mechanics seemed the only possible explanation in the light of the evidence. But it was a most unhappy conclusion in the context of classical scientific tradition. If the only basis of atomic existence is the probability cloud, and atoms are surrounded by charge clouds corresponding to their various electrons, then complex molecules must likewise consist of interacting probability wave amplitudes. And since living cells are composed of just such molecules, it follows finally that we, the creatures upon whom has fallen the responsibility for observing this uncertain world, are likewise so constituted—probability clouds contemplating other probability clouds in an ephemeral universe.

The philosophical implications have not been easy to accept, and are still being debated. The classical world was our oyster, and, like an old pair of shoes, ever so much more comfortable. It was nice to *know* things, or at least to think that one knew. For those whose education and awareness began in a tightly structured world it represents a devastating experience. One's reaction to such loosening of the nuts and bolts which hold the universe together is apt to be first disbelief and then despair. A generation reared in the mechanistic tradition of "Be good and you'll be rewarded, be bad and you'll be punished," grows up to conclude that it has been lied to; things are not like that at all. Suddenly there are no rules—or worse yet, someone keeps changing them as the game progresses.

And so we wander through "life's uncertain voyage" complaining bitterly of the unfairness of it all. Since we have no absolute hold over the future, cannot even know all of the present, what is the point of the struggle? There is nothing to do but float on the waves, listen to the sounds, watch the colors, do your thing. For how can one hope to control a destiny which hangs on a game of dice?

And yet the truth is that this world of quantum mechanics is not without structure. The determinism has merely shifted from the events themselves to the probability amplitudes which guide them. For the Schrödinger equation offers exact knowledge of the wave function, and the future behavior of the probability waves is completely determined by their past, just as in the classical world the events themselves were predetermined. Thus the quantum mechanical world provides an *exact* prediction of the *probability* of future events, despite the uncertainty of the events themselves. We have no way of knowing which radioactive atom will decay first; but we know how many of a large ensemble will decay in the course of an hour. We cannot know whether a piece of equipment sent to a distant planet will continue to function, but we know how to influence its chances. And when a living organism is irradiated at a particular exposure level, we do not know whether it will live or die; but we can predict quite accurately how many of a large group of such organisms will survive.

The inhabitants of this world may thus be assured, "to ease them of their griefs, their fears of hostile strokes, their aches, losses, their pangs of love, with other incident throes . . . ," that a new causality has emerged from the ashes of the old. The bewilderment of the physicists when they first discovered the quantum mechanical nature of the universe is matched by the feeling of strangeness experienced by a generation thrust into a world of modern scientific reasoning and unrestricted freedom of thought. There is the inevitable period of adjustment and experimentation while one discovers the rules of the game.

Actually there is much to recommend the universe in which we find ourselves. It may indeed be the best of all possible *physical* worlds (even though its *inhabitants* are still far from perfect). Planck's constant is *small* enough to provide the security of an environment which in most things appears to behave predictably.

On the other hand it is *large* enough to prevent atoms from collapsing and therefore to support the existence of the creatures who ask all the questions.

George Gamow in *Mr. Tompkins in Wonderland* has described an alternative world in which Planck's constant is so large as to affect the behavior of macroscopic objects. On a hunt it becomes difficult for the hunters to shoot a tiger which attacks them, because instead of occupying a localized region of space like ordinary earth tigers, it is instead spread out in a probabilistic "tiger cloud" as the result of the macroscopic uncertainty in its position. It is necessary to shoot the tiger "wave" many times, until a lucky shot finally hits it and forces it to drop like a "particle."

In the classical world it was possible to sit in a moving car without ambiguity. But in the strictly probabilistic world of quantum mechanics, although for the most part you are still sitting in the same car, a little bit of you leaks out into the field seen through the window, and an extremely small portion is two million light years away in the Great Nebula of Andromeda. Thus when your traveling companion complains, "You're not listening. Your mind is a thousand miles away," there is more truth in the statement than she realizes. This has in fact already passed beyond the whimsical stage, and is a part of current technology; the *tunnel diode* is a device which operates on this very principle, namely, that particles may be found on the other side of a barrier which in the classical sense it is impossible for them to penetrate.

The Schrödinger equation, like all theoretical models, of course has its limitations. One of these is that in its initial formulation it neglects the effect of relativity, describing the behavior of particles moving at relatively low speeds compared to that of light (which applies approximately to the atomic electrons). The postulates of quantum mechanics have, however, been combined with those of special relativity by Dirac, resulting in a relativistic quantum mechanics which is more appropriate to high energy processes. A dramatic result of this development was its theoretical prediction of the existence of antiparticles even before they were discovered. The birth and death of particles in our laboratories have motivated a further development known as quantum field theory, in which the processes of creation and annihilation of particles are incorporated into the mathematical formalism; here too the agreement between theory and experiment has been remarkable.

But despite the consistent successes of these extensions of quantum mechanics, the theories have always been plagued by mathematical difficulties. Even the Schrödinger equation, whose solution is so neatly obtained for the stationary states of the one-electron atom, requires approximation techniques in the case of the heavier atoms, and becomes inordinately difficult to solve for complex molecules. Much of the theoretical work has centered on developing a host of ingenious schemes for pushing forward the frontiers of successful solution, and applying them to physics, chemistry, and more recently to molecular biology. In principle, if we were only clever enough, there is no known reason why the theoretical models could not successfully explain the behavior of the large interacting systems of atoms and molecules which comprise life forms. There may be many surprises in store for us if we can manage to extend our physico-chemical analysis into other fields.

On the other hand the philosophical questions have not been fully resolved to everyone's satisfaction. Are the probability waves indeed to be taken literally in describing the behavior of individual systems? Or does the probabilistic interpretation apply only to large statistical ensembles? Physicists like Einstein and Bohm have continued to affirm their belief in an *a priori* material world which may be described independently of the act of observation. There is of course the operational problem of how to define such a world. Suggestions have been made that there may be "hidden parameters" (unobserved properties) which could restore determinism and causality. The fact that we have so far failed to find them, the argument goes, is a failure of the physicists and not of the world.

Various paradoxes have been proposed to dramatize the conceptual difficulties posed by quantum mechanics. In one of these, called "Schrödinger's cat," it is postulated that a cat has been confined to a steel chamber with a "hellish contraption" consisting of a very small amount of radioactive substance—so small in fact that there is a fifty percent probability that one of the atoms will decay in the course of an hour, and fifty percent that none will decay. If an atom decays, the emitted radiation triggers a counter which activates a hammer which smashes a container of cyanide, killing the cat. The wave function which describes the atom of course accurately predicts its behavior in the probabilistic sense. Like a particle which is passing partly through one hole of a barrier and partly through another, the final condition of the atom is a "mixture" of states, including both decay and failure to decay. This is all in accord with the successful treatment by quantum mechanics of such submicroscopic systems. But what about the cat? If the Schrödinger equation were to be solved for a system including not only the radioactive material, but also the "hellish contraption" and the cat, such a wave function would of necessity be a mixture of a live cat and a dead cat. Can the probability "cat cloud" include a superposition of a live and a dead cat? If a particle can be in two places at the same time, can a cat be both alive and dead?

Such paradoxes have been resolved to the satisfaction of most people, although they are revived periodically. The act of observation invariably discovers either the live cat or the dead cat, never both, just as the observation of the particle catches it in one hole or the other, never both. But it is somewhat more difficult to accept the effect of the interaction between the observer and a macroscopic cat.

The important question remains, how far can we go? While the high energy physicists probe the elementary particles in one direction, the biophysicists and microbiologists push the frontiers in the other. Is there a fundamental limit on the subdivision of particles into "lesser" particles, or do we get back to the same particles with which we started? Is there in fact such a thing as an *elementary* particle?

Does the uncertainty principle preclude knowledge about the nature of the *life* process? It could turn out that, like a slide prepared for observation by staining, live tissue is destroyed by the very process of understanding it. Can physico-chemical principles really provide a description of living organisms? Or is the knowledge that a cell is *alive* "complementary" to a complete description of its structure, as the particle is complementary to the wave?

What does quantum mechanics do to the concept of free will? Now that we have been unshackled from the constraints of a causal-deterministic structure, has the situation changed? Does "free" mean uncertain and unpredictable in the sense of being erratic? On the other hand, if the will is "rational" in being linked causally to its environment, what does it mean to say that it is free?

What about psychology? Even if we succeed in encompassing biological structures in the physico-chemical description, how can we describe the mind, a system in which the observer and that which is observed can be one and the same thing? Is it possible to understand the process of understanding itself?

And so the questions go on. But we shall surely not resolve them here, for this journey is ended.

Epilogue

This book is an experiment. A tradition has been violated: Physicists have for decades enjoyed a privileged position in society, and one of the prices has been preservation of a dichotomy between physics and nonscientific subjects. The benefits are obvious; not only have the physicists been generously supported, but the science itself has had a chance to develop in a relatively objective and detached environment, free of the vagaries of political fortune and the morass of unsolved social problems. Behind its protective wall physical science has thus managed to achieve a level of skill and sophistication which is unique in our civilization. Like many other important structures, however, this one too is beginning to show signs of cracking. The golden age of physics appears to be over, when society was willing to support the physicists while they talked only to each other. People today seem less sympathetic to the argument, "What I'm doing is important, but unfortunately I can't explain it to you because it's too complicated, and you don't know enough mathematics. But I assure you it is most important indeed."

If it is true (and not just a rationalization) that what happens in physics is important to everyone, then the connection has to be made. I believe that from now on a part of physics is going to be sharing it with other people—just as the affluent society sticks in our throats as long as it remains superimposed on a world in which so many people are ignorant and impoverished.

The dialogues between the poet and the scientist, as well as those of the other protagonists, have been deliberately polarized to provide a vehicle for some of the conflicting ideas heard on the current scene. They constitute a sort of argumentative therapy, and I have tried to make each position as convincing as I could regardless of my own viewpoint. Thus even the statements of the scientist are kept distinct from those of the author. There is no pretense to the delineation of real character. The poet is actually a bit humorless for my taste; I believe also that real poets have a little of the physicist in them, and it would be a sad world if none of the physicists had any poetry. I apologize to readers who identify with poets if they feel the presentation has been unfair or unsymmetric. This is after all a book about physics, and poets are quite eloquently having their day in court in most of the literature which appears nowadays.

Some readers of the preliminary manuscript have asked if a sort of agreement would not eventually be reached between the poet and the scientist; after all, both are intelligent people. This would, however, remind me too much of the statement released by politicians who have finished a conference and have not the slightest intention of telling anyone to what it is that they have agreed or failed to agree, assuming that they know themselves. It is precisely in the clash of opposing ideas that truth can eventually be found. The conflict between poet and scientist will not be resolved quickly, any more than the other problems of our times. The important thing is to keep the kettle boiling. A "united" people may be ideal for fighting old-time wars (against other "united" people), but consensus is rarely the mark of a free society.

There is an important distinction to be made between intuitive knowledge and scientific reasoning. Intuition reminds us of things which on some level we have long known and are otherwise likely to forget. But scientific reasoning tells us things we have never known or suspected; it may even contradict what we always thought we knew. It is a mistake to believe that the only way to cultivate one of these faculties is to reject the other.

If I had an incurable disease with only a year to live, would I want to know? One person may say, "What good is it to know? It will only ruin the remaining time." The other says, "But I am not an unreasoning creature. A human being deserves to *know* things. If the information is available, and particularly if others know, the very act of concealment destroys a part of me in advance, for I am then something less than those who do know." A man with a whole mind wants to know everything he can, and is not content to play the old records till the day he dies.

The understanding of atomic structure gained from quantum mechanics has changed the character of physics, transformed chemistry from a phenomenology into an analytic science, and is making rapid inroads into biology. Now biochemical processes can be modeled and analyzed on the molecular level, and we find ourselves at the threshold of the secrets of heredity and life. It ought to be a good time to be alive; yet large sections of the youth are suffering a profound malaise, and everyone is trying to understand the basis for the phenomenon of rejection.

For centuries educated men have struggled to defeat poverty, ignorance, and superstition; now there comes a generation which questions the value of the struggle. Every age has proudly passed its discoveries on to the next, which seized them eagerly even while chafing at the confinements and restrictions of the past; now comes an affluent and relatively educated young population which declines to accept the torch. "What was so bad about the lot of the ignorant savage?" they say. "His world was no more absurd than yours; his reality no less real than yours; and the feeling of security he gave his children must have been greater than that which you have given us."

And so the primitive medicine man is studied like a sage, and "brain stimulation" is proposed as a substitute for the external world. I do not really believe that in stopping to ask such questions they are embracing the old paths of ignorance and superstition, although like other generations this one too will have its casualties. It is just that scientific progress has been so rapid that people want to pause for breath.

I have tried to suggest that one of the things to be learned from the developments of physics is that the truth can have many faces and speak with many voices. The important thing is that all the faces be seen and none of the voices be stilled. If this is done the problems of society will be solved—provided of course they do not first reach the irreversible stage in which the nuclear-tipped missiles make their way out of the silos and the seas and take the world away from the people.

A generation has been born many of whose members hold the scientists responsible for acts committed with the results of science, and which refuses to be convinced by the old argument, "That's not my department." The world appears to them to make no sense at all, and the existing power structure to be something out of Lewis Carroll. I expect that the basic problems of our time are going to revert finally to this generation—those born in the nuclear age. For the older generation is still focusing on the problems of the last decade and trying to solve them by the methods of the last century. The young people do not suffer this handicap, but they cannot live and work without knowledge; and when they choose to reject the traditions of the past, I hope they will not reject also the heritage of reason and knowledge which is properly theirs. One can at least ask that before surrendering this birthright they give themselves a chance to know what is being given up. For it would be a pity to have to start all over again, and even the young grow old too quickly.

Appendix

The background required for reading this section is that of high school mathematics. The relativity development presupposes only intermediate algebra, and the optical equations require a knowledge of the trigonometric functions (sine, cosine, tangent).

The reader is, however, warned to expect at this point a certain change of pace. If mathematics is for him a foreign language (as it is, alas, for so many), he may find himself stumbling on occasion. Since all of us are foreigners somewhere, this should not be an unfamiliar experience; a small assist from a "native" in such cases will go a long way.

The derivations given below are not absolutely essential to an appreciation of the concepts which have been introduced, if one is willing to accept a few mathematical results. On the other hand, a small amount of coping with the language barrier is most rewarding and opens many doors to understanding.

Chapter 6

First we deduce the *Lorentz transformation*, which relates the position and time of an event in one frame of reference to the position and time of the same event observed in another frame traveling at constant velocity with respect to the first. The "event" could be, for example, the presence of an object like racing car 2 of Fig. 6–1 in a particular position (labeled P in

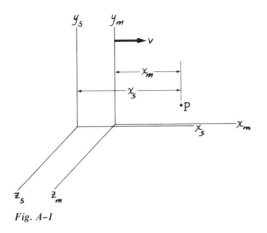

Fig. A–1

Fig. A–1) at a particular time. The frame of reference is designated by a set of coordinate axes. We may define the origin of coordinates wherever we like in a given frame of reference, and can call any direction we wish the x-direction. Let us consider the first frame to be "stationary" (the event is labeled by space-time coordinates x_s, y_s, z_s, t_s), and the second frame (same event labeled by coordinates x_m, y_m, z_m, t_m) to be moving in the x-direction (to the right) with relative velocity v. At time designated as zero ($t_m = t_s = 0$) the two frames coincide ($x_m = x_s$, and likewise for the other two coordinates), but the moving frame continues its motion at constant velocity v to the right. The labels "moving" and "stationary" are of course only a matter of convenience. Someone could with equal justification call the second frame the "stationary" one, and consider the first frame to be moving to the left. We are here concerned with only the one spatial coordinate x, although the treatment is easily extended to include y and z.

Classically (before relativity) the space-time coordinates describing an event were related in these two frames by what is known as the *Galilean transformation* of coordinates, namely,

$$x_m = x_s - vt_s \, , \tag{A-1}$$
$$t_m = t_s \, . \tag{A-2}$$

Equation (A–1) is an expression of the fact that when an object is at rest in the moving frame, it must be moving to the right with velocity v in the stationary frame. (The easiest way to see this is to consider an example and substitute numerical values in the equation.) Equation (A–2) is simply the classical assumption of absolute and common time for observers in the two frames; it appears in fact pointless to use different labels for t_m and t_s, but we do so in preparation for what follows. For completeness we should also add that $y_m = y_s$ and $z_m = z_s$, but these coordinates do not concern us here, and we may take them to be zero.

Consider now an object which starts out at the origin of coordinates at time zero and moves with constant velocity in the x-direction. (Note that the velocity of the object is not in general equal to the velocity v of the moving frame.) At some later time t_s it reaches a position which in the stationary frame has coordinate x_s. Since it has traveled a distance corresponding to x_s in time t_s, its velocity V_s (which is distance divided by time) must, according to an observer using the stationary frame, be

$$V_s = x_s/t_s \ .$$

The velocity of this same object as observed from the moving frame is

$$V_m = x_m/t_m \ .$$

To get the relationship between V_s and V_m, we divide each side of equation (A–1) by the corresponding side of (A–2), obtaining

$$\frac{x_m}{t_m} = \frac{x_s}{t_s} - v \ , \tag{A-3}$$

or

$$V_m = V_s - v \ . \tag{A-4}$$

Equation (A–4) corresponds exactly to equation (6–1), except that here we are using the symbols V_m, V_s, and v in place of u, v_2, and v_1, respectively. Our stationary frame corresponds to that of the spectators in the stand in Chapter 6, and the moving frame is tied to car 1. The "object" whose motion is being described could be car 2 (or its driver).

Equation (A–4) is the old rule for addition (or subtraction) of velocities, the rule which broke down as a result of the Michelson–Morley experiment. To derive the simplest possible new rule consistent with the experimental results, we assume that equations (A–1) and (A–2), which are "linear" in the sense that they contain no higher-degree terms like t_s^2 or x_s^3, will be replaced in the new transformation by another pair of similarly linear equations.

The most general linear replacement for equation (A–1) which still satisfies the requirement that a particle at rest in the moving frame moves at velocity v in the stationary frame (we want to make as few changes in the "old" physics as possible) is

$$x_m = \gamma(x_s - vt_s) , \qquad (A-5)$$

where γ (gamma) is an as yet unknown constant. That this is the most general replacement of equation (A–1) may not be immediately obvious. One might, for example, argue that it ought rather to be

$$x_m = \gamma x_s + \alpha t_s , \qquad (A-5a)$$

where γ and α are constants applying to all observed events. But consider now the special case of a particle at rest in the moving frame ($x_m = 0$); in the stationary frame it must have velocity

$$V_s = v = x_s/t_s .$$

Substituting therefore $x_s = vt_s$ in equation (A–5a), we have, since $x_m = 0$,

$$0 = \gamma vt_s + \alpha t_s ,$$

whence

$$\alpha = -\gamma v ,$$

resulting in equation (A–5).

The most general linear equation to replace equation (A–2) is

$$t_m = Ax_s + Bt_s , \qquad (A-6)$$

where A and B are as yet unknown constants.

In the former Galilean transformation both γ and B were equal to one, and A was zero; we must discover what they should be to agree with the results of the Michelson–Morley experiment.

Suppose that a light source at the origin of coordinates is turned on at time zero, when the two frames coincide. Then in the stationary frame at a later time t_s, the light signal will have traveled a distance $|x_s|$ in both the positive and negative x-directions, such that

$$|x_s| = ct_s . \qquad (A-7)$$

By the same token this signal may be described in the moving frame, where it will have traveled a distance $|x_m|$, such that

$$|x_m| = ct_m . \qquad (A-8)$$

Thus in either frame and in both directions, the distance traveled by the light signal equals its velocity c times the time of travel. We are here using the Michelson–Morley result that the speed of light, c, is the same in all frames. (Otherwise we should have used c_s in the first case and c_m in the second, to distinguish them.) The absolute value sign | | merely expresses the fact that, regardless of whether x_s and x_m are positive or negative quantities, time t_s or t_m is taken to be positive.

Both sides of equations (A–7) and (A–8) are now squared and rewritten

$$x_s^2 - c^2 t_s^2 = 0 \ , \tag{A–9}$$

$$x_m^2 - c^2 t_m^2 = 0 \ . \tag{A–10}$$

Since the left-hand sides of (A–9) and (A–10) are both equal to zero, they can be set equal to each other,

$$x_m^2 - c^2 t_m^2 = x_s^2 - c^2 t_s^2 \ . \tag{A–11}$$

Now substitute (A–5) and (A–6) in (A–11):

$$\gamma^2(x_s^2 - 2vx_s t_s + v^2 t_s^2) - c^2(A^2 x_s^2 + 2ABx_s t_s + B^2 t_s^2) = x_s^2 - c^2 t_s^2 \ .$$

This may be rewritten

$$(\gamma^2 - A^2 c^2)x_s^2 - 2(\gamma^2 v + ABc^2)x_s t_s + (\gamma^2 v^2 - c^2 B^2)t_s^2 = x_s^2 - c^2 t_s^2 \cdot \tag{A–12}$$

Now there is an algebraic principle which states that if x_s and t_s are independent variables, then an equation like (A–12) will be satisfied for *all* possible values of x_s and t_s only if the corresponding coefficients of x_s^2, t_s^2, and $x_s t_s$, on both sides of the equation are equal. In other words,

$$\gamma^2 - A^2 c^2 = 1 \ , \tag{A–13}$$

$$\gamma^2 v^2 - c^2 B^2 = -c^2 \ , \tag{A–14}$$

and

$$-2(\gamma^2 v + ABc^2) = 0 \ . \tag{A–15}$$

From (A–13),

$$A^2 = \frac{\gamma^2 - 1}{c^2} \ . \tag{A–16}$$

From (A–14),

$$B^2 = \frac{\gamma^2 v^2 + c^2}{c^2} \ . \tag{A–17}$$

Substituting (A–16) and (A–17) in (A–15), we have

$$-\gamma^2 v = \sqrt{(\gamma^2 - 1)(\gamma^2 v^2 + c^2)} \ .$$

Squaring both sides, we obtain

$$\gamma^4 v^2 = \gamma^4 v^2 + \gamma^2 c^2 - \gamma^2 v^2 - c^2$$

This may be solved for γ, giving

$$\gamma = \frac{1}{\sqrt{1 - v^2/c^2}} \cdot \tag{A-18}$$

The positive value of the square root in this expression is the one which applies, since for velocities much less than the speed of light, i.e., as v/c approaches zero, γ must approach $+1$, making (A–5) consistent with (A–1) of the Galilean transformation, and agreeing with previous experience at "ordinary" velocities.

When this value of γ from (A–18) is now substituted into (A–17), we get (after algebraic manipulation)

$$B = \frac{1}{\sqrt{1 - v^2/c^2}} = \gamma . \tag{A-19}$$

(Again the positive root guarantees $B \to +1$ as $v/c \to 0$.)

Substituting (A–18) into (A–16) gives

$$A = -\frac{1}{\sqrt{1 - v^2/c^2}} v/c^2 = -\gamma v/c^2 . \tag{A-20}$$

[The minus sign in (A–20) is due to the fact that this time the negative value must be taken for the square root of A^2, since this is the only way to satisfy (A–15) for positive B.]

When (A–18), (A–19), and (A–20) are substituted in (A–5) and (A–6), we get the *Lorentz transformation,*

$$x_m = \frac{1}{\sqrt{1 - v^2/c^2}} (x_s - vt_s) , \tag{A-21}$$

$$t_m = \frac{1}{\sqrt{1 - v^2/c^2}} (t_s - vx_s/c^2) . \tag{A-22}$$

(If we had considered all three spatial dimensions instead of only one, we would have also $y_m = y_s$ and $z_m = z_s$, thus completing the set of four equations involving three spatial and one time dimension.)

To get Einstein's new rule for addition of velocities, we note that the velocity V_m observed in the moving frame is distance divided by time (also in the moving frame),

$$V_m = \frac{x_m}{t_m} = \frac{x_s - vt_s}{t_s - vx_s/c^2} ,$$

where (A–21) and (A–22) have been utilized. Dividing both numerator and denominator of the right-hand side of this expression by t_s, we get

$$V_m = \frac{(x_s/t_s) - v}{1 - (x_s/t_s)(v/c^2)} \, ;$$

or, since

$$\frac{x_s}{t_s} = V_s \, ,$$

$$V_m = \frac{V_s - v}{1 - V_s v/c^2} \, . \tag{A–23}$$

It may be seen that (A–23) approaches (A–4) for small velocities, i.e., when $v/c \to 0$, as it should.

We can also turn equation (A–23) inside out by solving it for V_s in terms of V_m, getting

$$V_s = \frac{V_m + v}{1 + V_m v/c^2} \, . \tag{A–24}$$

This is the Einstein law of addition of velocities, which replaced the old addition rule $V_s = V_m + v$. Its derivation incorporates the postulates of special relativity, and therefore it is consistent with the findings of Michelson and Morley. Furthermore, when we let $V_s = V_m = c$, corresponding to replacing driver 2 with a light signal, and allow driver 1 to approach the speed of light ($v \to c$), then equation (A–24) becomes

$$c = \frac{c + c}{1 + c^2/c^2} \, ,$$

which is a mathematically consistent expression, instead of the nonsensical result of equation (6–4).

Chapter 8

To derive the expression for contraction of lengths, we consider a rod (lined up in the x-direction) which is at rest in the moving frame. In this moving frame, which has velocity v with respect to the stationary frame, the rod has its "rest length" L_0. The question is, what is the length L of the rod as determined by an observer using the *stationary* frame?

Each observer measures the length of a rod by determining at a particular instant of time the positions of the two ends of the rod in his frame of reference (Fig. A–2). If the left end of the rod is designated by subscript

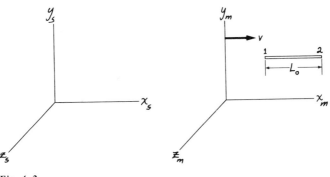

Fig. A-2

1 and the right end by 2, then the position of either end of the rod in the moving frame is related to its position in the stationary frame, according to equation (A-21), by

$$x_{1m} = \frac{1}{\sqrt{1 - v^2/c^2}} (x_{1s} - vt_s) , \tag{A-25}$$

and

$$x_{2m} = \frac{1}{\sqrt{1 - v^2/c^2}} (x_{2s} - vt_s) , \tag{A-26}$$

where t_s is the precise instant at which the two ends are observed in the stationary frame. (It is, of course, important that the ends of the rod be observed *simultaneously* when its length is measured, since this is what we mean by "length"; otherwise the measurement will not be valid for a moving rod.)

The length of the rod in the moving frame is defined by

$$x_{2m} - x_{1m} = L_0 , \tag{A-27}$$

and in the stationary frame by

$$x_{2s} - x_{1s} = L . \tag{A-28}$$

Substituting equations (A-25) and (A-26) in (A-27) yields

$$\frac{1}{\sqrt{1 - v^2/c^2}} (x_{2s} - vt_s) - \frac{1}{\sqrt{1 - v^2/c^2}} (x_{1s} - vt_s) = L_0 .$$

Since vt_s cancels in this expression,

$$\frac{1}{\sqrt{1 - v^2/c^2}} (x_{2s} - x_{1s}) = L_0 .$$

Substituting (A–28), we have

$$\frac{1}{\sqrt{1 - v^2/c^2}} L = L_0 \ ,$$

whence

$$L = L_0 \sqrt{1 - v^2/c^2} \ . \tag{A–29}$$

This is the *Lorentz-Fitzgerald contraction* stated in equation (8–1).

The time dilation equation (8–2) can be obtained by first inverting (A–21) and (A–22), that is, solving them simultaneously for x_s and t_s in terms of x_m and t_m. When this is done, the inverse of (A–22) is found to be

$$t_s = \frac{1}{\sqrt{1 - v^2/c^2}} (t_m + vx_m/c^2) \ . \tag{A–30}$$

We consider the elapsed time interval measured on a clock which is at rest in the moving frame, and hence moving at velocity v in the stationary frame. To an observer (the R.O., for example) who is using the moving frame in which this (rocket) clock is at rest, the time interval between two ticks of the clock is defined to be

$$T_0 = t_{2m} - t_{1m} \ , \tag{A–31}$$

where t_{1m} and t_{2m} are the times at which these ticks occur according to an observer using the moving frame. What we wish to find is the corresponding time interval

$$T = t_{2s} - t_{1s} \tag{A–32}$$

between the same two ticks of this (rocket) clock as measured by an observer (the P.O.) who uses the stationary frame.

From equation (A–30),

$$t_{1s} = \frac{1}{\sqrt{1 - v^2/c^2}} (t_{1m} + vx_{1m}/c^2) \ , \tag{A–33}$$

$$t_{2s} = \frac{1}{\sqrt{1 - v^2/c^2}} (t_{2m} + vx_{2m}/c^2) \ , \tag{A–34}$$

where x_{1m} and x_{2m} are the locations of the clock in the moving frame at times t_{1m} and t_{2m}, respectively. Subtracting (A–33) from (A–34), we have

$$t_{2s} - t_{1s} = \frac{1}{\sqrt{1 - v^2/c^2}} (t_{2m} + vx_{2m}/c^2 - t_{1m} - vx_{1m}/c^2) \ . \tag{A–35}$$

But since the clock is at rest in the moving frame, its position remains unchanged in this frame, that is, $x_{1m} = x_{2m}$.

Therefore these position terms drop out of (A–35). If we now also substitute the definitions (A–31) and (A–32) in (A–35), the latter becomes

$$T = \frac{T_0}{\sqrt{1 - v^2/c^2}}, \tag{A–36}$$

the time dilation equation (8–2).

Chapter 9

Einstein obtained equation (9–2) from Maxwell's equations. This undoubtedly contributed to his deduction that light comes in energy bundles, or *quanta*, whose magnitude is proportional to the frequency of the oscillating source. (See Chapter 14.) We shall, however, do it the other way around, by utilizing this frequency-energy proportionality and also the time dilation equation (8–2) or (A–36). The reader should therefore have read up to (and including) Chapter 14 before attempting to understand this derivation.

Let us first replace the cheap flashlights by monochromatic sources of light; the emitted energy E in the platform frame of reference is necessarily proportional to the frequency ν, in accordance with equation (14–7). Now we know that people riding the rocket do not observe this same frequency of light as those riding the platform. Actually there are two reasons for this: The first is that all clocks on the platform (which the rocket people consider to be moving with a velocity of magnitude v relative to themselves) are observed by them to run slow; and a frequency source is a clock, since it sends out a fixed number of wave cycles per second, which can be counted to measure time. Because the time measured by the rocket people between two "ticks" (or wave peaks) of such a clock is *greater* by the factor $1/\sqrt{1 - v^2/c^2}$ as a result of this clock's relative motion, its frequency is *decreased* by the same factor; namely, the slowed-down clock is observed to have, instead of frequency ν, a smaller frequency $\nu\sqrt{1 - (v^2/c^2)}$.

But in addition to this slowing down of the frequency source as a result of time dilation, there is also a Doppler shift in the frequency which is received, rather like the change in pitch of sound waves coming from a siren on a moving vehicle. (This effect exists even without relativity.) We are here assuming that the rocket comes so close to the platform that the line between a pair of rocket observers stationed along the path of relative motion passes through the two flashlights. (Perhaps it is best to picture the rocket as passing right *through* the platform.) Thus one of the flashlights is directly approaching a rocket observer and the other is receding from a (different) rocket observer. Now when a wave source approaches an observer at velocity v, the emitted waves are caused to bunch up with respect to previously emitted waves as a result of this forward motion of the source, producing an effective decrease in wavelength. The length of wave emitted during a single period T, instead of the value λ it has when the source is at rest, becomes $\lambda - vT$, where the period T of the wave is multiplied by the velocity v of the source to obtain the distance vT by which the wavelength has been foreshortened. The factor by which the wavelength is diminished as a result of this source motion is therefore

$$\frac{\lambda - vT}{\lambda} = 1 - \frac{vT}{\lambda} = 1 - \frac{v}{c},$$

from equation (12-4). But this *decrease* in wavelength must correspond, from equation (13-2), to a reciprocal *increase* of $1/(1 - v/c)$ in the frequency, since, from equation (13-2), a decrease in wavelength λ is possible only with a corresponding increase in frequency v (the speed of light c remaining constant).

When this Doppler effect is combined with that of time dilation, the observer on the rocket, instead of receiving frequency v (which he would if the source were at rest relative to himself), observes a frequency

$$v_1 = v \frac{\sqrt{1 - v^2/c^2}}{1 - v/c}$$

for the light waves which are approaching him, i.e., traveling in the *same* direction as the platform source with respect to this rocket observer. (This applies to the light emitted by only one of the two flashlights. The other flashlight sends out light in a direction *opposite* to that of its relative motion with respect to a rocket observer.) Since the emitted energy is proportional to the frequency, the amounts of light energy observed in the two frames of reference have the same relationship as the frequencies. The energy $E/2$ emitted in a given time by this first flashlight (half the total energy E emitted by both flashlights) according to a platform observer, therefore becomes for the rocket observer

$$E_1 = \frac{E\sqrt{1 - v^2/c^2}}{2(1 - v/c)}.$$

The second flashlight, on the other hand, emits light waves in a direction opposite to that of its own motion. The Doppler shift factor for this flashlight therefore becomes $1/(1 + v/c)$, corresponding to a *decrease* of frequency; hence this energy as observed in the rocket frame of reference becomes

$$E_2 = \frac{E\sqrt{1 - v^2/c^2}}{2(1 + v/c)} .$$

The total energy E' emitted by the two flashlights as observed in the rocket frame of reference is the sum of these two contributions. Thus

$$E' = E_1 + E_2 = \frac{E}{2} \sqrt{1 - \frac{v^2}{c^2}} \left(\frac{1}{1 - v/c} + \frac{1}{1 + v/c} \right) .$$

When this expression is put over a common denominator and reduced algebraically, it becomes

$$E' = \frac{E}{\sqrt{1 - v^2/c^2}} , \tag{A-37}$$

corresponding to equation (9–2). Of course an ordinary flashlight is a thermal source of energy, consisting of many frequencies, not just one, like a monochromatic source; but all these frequencies and their corresponding energies increase in the same proportion, so that the above result holds also for ordinary light sources.

We next carry out the algebraic steps in going from equation (9–4) to (9–5):

$$\frac{E}{\sqrt{1 - v^2/c^2}} - E = \tfrac{1}{2}mv^2. \tag{9-4}$$

The square root factor in the denominator of the first term of (9–4) may be expanded by means of the binomial theorem,

$$\frac{1}{\sqrt{1 - v^2/c^2}} = (1 - v^2/c^2)^{-1/2} = 1 + \tfrac{1}{2}v^2/c^2 + \cdots , \tag{A-38}$$

where higher-order terms in v^2/c^2 may be neglected because it is such a small quantity. Equation (9–4) therefore becomes

$$E(1 + \tfrac{1}{2}v^2/c^2 - 1) = \tfrac{1}{2}mv^2 , \tag{A-39}$$

thus allowing the cancellation of the ones, and then of the v^2, and thereby leading to the famous result of equation (9–5), $E = mc^2$. This relationship gives the energy contained in a mass m as observed in a frame of reference in which it is at rest, since E was defined in the platform frame in which the flashlight batteries are at rest. It is therefore called the *rest energy*.

This predicted mass-energy equivalence is the direct result of Einstein's simple explanation of the failure to detect an ether wind, namely, that the laws of nature (and the speed of light) are the same in all inertial frames. Previously one would have attributed the additional light energy in the "moving" frame to the effect of an ether wind.

Chapter 13

Interference fringes can be obtained at the screen when parallel rays of coherent monochromatic light pass through the two slits A and B in the barrier (Fig. A–3). The distance d between slits A and B should be larger than the wavelength λ of the light, but much less than the distance f to the screen; also, the width of either slit should be much less than the wavelength.

The light emerging from the slits spreads out in all forward directions. (This is known as Huygens' principle.) We consider one particular direction corresponding to the angle θ.

The lines AD and BD' represent those parallel rays which emerge from the slits at angle θ. If the distance f to the screen were infinite, the points D and D' would be one and the same, since parallel lines meet at infinity; however, the same effect is obtained by introducing a lens (not shown) immediately after the barrier which actually does bring D and D' to a focus at their midpoint D''.

When the angle θ is zero the rays from the two slits arrive at the screen with light waves exactly in phase (because the distances of travel are equal); this results in reinforcement and hence produces a bright region called the *central maximum* at point O', which is directly opposite the midpoint O between the slits. But when θ is greater than zero (as in the diagram), the

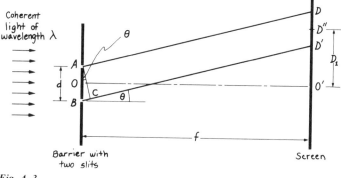

Fig. A–3

light emerging from the slits differs in phase an amount determined by the length of line BC, which is obtained by drawing AC perpendicular to BD'. From geometrical considerations,

$BC = d \sin \theta$.

Now there is a particular value of θ such that BC is equal to precisely half a wavelength; that is,

$$d \sin \theta = \lambda/2 . \tag{A-40}$$

At this angle the two emerging rays will be precisely out of phase, since a half-wavelength phase difference means that a wave crest at A corresponds to a trough at C, and vice versa. This is the angle at which the *first minimum* appears on the screen, since the two emerging rays then interfere destructively. There are, of course, minima at all the odd multiples of half a wavelength,

$$d \sin \theta = n\lambda/2 , \tag{A-41}$$

where $n = 1, 3, 5, 7, \ldots$; but we are here concerned only with the first.

The distance D_1 from the central maximum to the first minimum is, because of the geometry,

$$D_1 = f \tan \theta . \tag{A-42}$$

Now we have specified that the slit spacing d must be larger than the wavelength. In actual experiments, d is for practical reasons usually a great deal larger than the wavelength λ of the light, which is measured in Angstrom. This simplifies expression (A–42), since (A–40) may be written

$$\sin \theta = \lambda/2d , \tag{A-43}$$

which is a very small quantity; and for such small angles the sine of the angle is effectively equal to the tangent. We may therefore replace the tangent in (A–42) with the sine, obtaining, when it is combined with (A–43),

$$D_1 = \lambda f/2d , \tag{A-44}$$

which is just equation (13–1). When there is no lens for converging the rays (as in the original form of Young's experiment) the same result is obtained, but with a few added steps in the derivation.

Chapter 14

To show that the mass of the photon is zero, we utilize the results of Chapter 9. According to equation (9–5), the energy E emitted by the flashlights as observed in a frame of reference in which they are at rest is related to the amount of mass m which is converted into energy by

$$E = mc^2 . \tag{A-45}$$

In the "moving" rocket frame, however, this energy is instead observed to be, according to equation (9-2),

$$E' = \frac{E}{\sqrt{1 - v^2/c^2}} . \tag{A-46}$$

Substituting (A-45) into (A-46), we see that the energy E' which is released by the conversion of an amount of mass m, as observed in a frame in which the mass is moving with velocity v, is given by

$$E' = \frac{mc^2}{\sqrt{1 - v^2/c^2}} . \tag{A-47}$$

Whereas (A-45) gives the mass-energy conversion relationship for a mass which is at rest with respect to the observer, (A-47) gives the corresponding relationship when the mass is moving with velocity v. Suppose now that this mass m, instead of being a part of the flashlight batteries, refers to a single particle which releases energy upon having its mass completely converted. The same reasoning which led to (A-45) and (A-47) still applies.

The quantity mc^2 of equation (A-45), as has been stated, is called the rest energy of the particle, since it is the energy into which it can be converted as seen in a frame in which it is at rest. In a frame in which the particle is moving, on the other hand, the energy E' is related to the mass m of the particle by (A-47), where v is the velocity of the particle.

Now if this velocity were to approach the speed of light c, then the denominator of (A-47) would approach zero. A vanishing denominator causes a fraction to become infinite unless the numerator also vanishes. Hence unless such a particle has infinite energy, the numerator of (A-47) must go to zero in this case, which requires the mass to be zero. This limiting condition applies to the photon, which is in fact a particle of light traveling at velocity c. Clearly its energy is not infinite in general; hence the mass of the photon must be zero.

The momentum defined by equation (14-5) is a quantity which was observed to be conserved in collisions at ordinary velocities like those of billiard balls, automobiles, etc. But for particles moving at relativistic velocities comparable to the speed of light, this quantity is found not to be conserved in general; what is instead observed (both theoretically and experimentally) to be conserved in high-velocity collisions is the quantity

$$p = \frac{mv}{\sqrt{1 - v^2/c^2}} \; . \tag{A-48}$$

This is therefore the more general definition of momentum, and may be seen to approach the old approximate definition of equation (14-5) at low velocities, where v is much smaller than c.

Suppose now in the case of a particle of mass m that we square both sides of (A-47) and subtract from it the square of the rest energy of the particle:

$$E'^2 - m^2c^4 = \frac{m^2c^4}{1 - v^2/c^2} - m^2c^4$$

$$= \frac{m^2c^4 - m^2c^4 + m^2v^2c^2}{1 - v^2/c^2}$$

$$= \frac{m^2v^2c^2}{1 - v^2/c^2} \; . \tag{A-49}$$

Substituting (A-48), we may write this

$$E'^2 - m^2c^4 = p^2c^2 \; . \tag{A-50}$$

If the relativistic particle in (A-50) happens to be a photon, its mass m is zero, and in such a case therefore

$$E' = pc \; , \tag{A-51}$$

which corresponds to equation (14-8), except that there we did not use a prime on the E; note that a photon is not at rest in any frame, but moves at the same speed c for *all* observers (the second postulate of special relativity).

Chapter 15

The relationship between the radius of a circular orbit and the speed of the revolving body follows from Newton's second law of motion, as applied to nonrelativistic particles of fixed mass m,

$$F = ma \; . \tag{A-52}$$

The radial acceleration of a circular orbit, in terms of the velocity v and the radius r, is

$$a = v^2/r \; . \tag{A-53}$$

For an earth satellite or a planetary electron revolving around the nucleus, the force F is inversely proportional to the square of the distance r from the center of force; namely,

$$F = K/r^2 ,\qquad\qquad (A\text{--}54)$$

where K is the constant of proportionality. (For earth satellites, K is the product of the universal gravitational constant, the mass of the satellite, and the mass of the earth; for the atom it is the product of the electron charge and the nuclear charge.)

Substituting (A–53) and (A–54) in (A–52), we have

$$K/r^2 = mv^2/r ,$$

or

$$v = \sqrt{K/mr} .\qquad\qquad (A\text{--}55)$$

Thus the radius of a circular orbit uniquely determines the speed, and vice versa.

De Broglie's explanation of equation (15–3) or (15–4) is obtained by fitting an integral number n of wavelengths λ into the circumference of a circular orbit. In other words,

$$n\lambda = 2\pi r ,\qquad\qquad (A\text{--}56)$$

where $n = 1, 2, 3, 4, \ldots$.

Substituting $\lambda = h/p$ in (A–56), we have

$$nh/p = 2\pi r ,$$

whence

$$pr = nh/2\pi ,\qquad\qquad (A\text{--}57)$$

corresponding to equation (15–4). Thus Bohr's postulate of equation (15–3) or (15–4), which he proposed as an *ad hoc* explanation of the spectroscopic data, may be *deduced* simply by fitting an integral number of de Broglie wavelengths into a circumference.

If we substitute (A–55) into (15–3), this gives the "allowed" radii of the Bohr atom,

$$r = n^2 h^2/4\pi^2 mK ,\qquad\qquad (A\text{--}58)$$

where $n = 1, 2, 3, \ldots$.

Chapter 16

Single-aperture diffraction may be analyzed to give the distance Δx between the central maximum and first minimum, by considering the two halves of the slit as interfering sources (Fig. A–4). (The analysis of a circular aperture of diameter D is somewhat more complicated than that of a slit of width D, but the result is not very different. For simplicity of calculation, therefore, we here replace the circular lens with a slit of the same dimension.)

Coherent monochromatic light passes through the single slit of width D, and rays emerge in all directions toward the screen. As in the case of two-hole interference, we consider a particular direction defined by angle θ. However, now the width D of the slit is much *larger* (instead of smaller) than the wavelength λ of the light. We consider two rays emerging from the edge A and the midpoint B of the slit. By the very same process as for two-slit interference (Appendix, Chapter 13), and corresponding to equation (A–40), these two rays interfere destructively to obtain a minimum at angle θ defined in this case by

$$(D/2) \sin \theta = \lambda/2 . \tag{A–59}$$

The two parallel interfering rays come to a focus at F because in a lens system the distance f is the focal length of the lens.

Now consider another pair of rays, also at angle θ, but emerging from two points slightly below A and B, respectively. These likewise interfere destructively when θ satisfies (A–59), since the same diagram is merely displaced downward slightly. This process is continued from the pair of points A and B to successively lower pairs of points in the slit until reaching B and C, respectively, thus sweeping through the entire slit. All rays emerging at angle θ interfere destructively in pairs, one ray of the pair from the upper half of the slit and the other from the lower half. Thus all the rays at angle θ interfere destructively when θ satisfies (A–59), and this defines

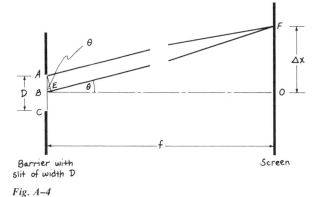

Fig. A–4

the first minimum. (As with two slits, there are also other minima at other appropriate values of θ.) The distance Δx from the central maximum (which is centered at O by the same reasoning as for two slits) to the first minimum satisfies

$$\tan \theta = \Delta x/f \,, \tag{A–60}$$

in accordance with the geometry of Fig. A–4. Equation (A–59) may be written

$$\sin \theta = \lambda/D \,, \tag{A–61}$$

which is a very small quantity, since we have specified that the width of the slit is much greater than the wavelength. This makes it possible to approximate $\sin \theta$ by $\tan \theta$, and hence combine (A–60) with (A–61), obtaining

$$\Delta x \approx \lambda f/D \,, \tag{A–62}$$

the result referred to in equation (16–1).

[The focal length f of a lens is actually not completely independent of its diameter D, and it may be more meaningful to rewrite this relationship after dividing both sides by f. The ratio $\Delta x/f$, which in accordance with equation (A–60) is the *angular resolution* of the lens, is the more fundamental quantity than Δx, and it is determined solely by the λ/D ratio.]

The uncertainty Δp in the extent of the momentum disturbance to the electron is due to our inability to know which part of the lens the photon passed through after collision with the electron. What we finally observe is of course the point of arrival of the photon, and what is uncertain is the angle ϕ in Fig. 16–1 at which it emerges from the lens. Actually it is the (vertical) x-component p_x of the emerging photon's momentum p which determines this angle (as shown in Fig. A–5). If the lens were of zero diameter, ϕ would be zero and there would be no uncertainty about p_x, which would be simply zero; it would then be possible to reconstruct the history of the collision and know exactly how much the electron was originally deflected. But with a nonzero-sized lens, p_x is an unknown quantity which reflects a corresponding uncertainty of the very same amount in our knowledge of the electron momentum. It is the maximum value of this quantity which must therefore be computed.

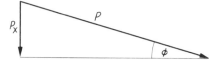

Fig. A–5

The diagram shows the photon's momentum vector of magnitude p, and its component p_x in the x-direction. For any angle ϕ,

$$\tan \phi = p_x/p = \lambda p_x/h , \tag{A-63}$$

where the de Broglie relation $\lambda = h/p$ has been utilized.

But it is also seen from the geometry of Fig. 16-1 that the *maximum* possible value of ϕ, corresponding to the photon's having passed through the outside edge of the lens, is given by

$$\tan \phi_{max} = D/2f . \tag{A-64}$$

This is the angle at which p_x has its maximum value and is therefore equal to the uncertainty Δp we are looking for. Combining (A-63) and (A-64), and replacing the maximum value of p_x by Δp, we obtain

$$\Delta p = (h/\lambda)(D/2f) , \tag{A-65}$$

which is just equation (16-2).

We now compute the momentum uncertainty for the electron in the ground state of the hydrogen atom as a result of observing its position accurately enough to "see" the orbit, as stated in equation (16-5). For the radius of the orbit we use the value predicted by Bohr's classical model, as stated in equation (15-4) for the first Bohr orbit. From this equation the radius is

$$r = h/2\pi p , \tag{A-66}$$

where p is the momentum of the electron in this orbit. (The radius in Bohr's "classical picture" agrees with the *most probable* radius predicted by the Schrödinger equation.)

Substituting (A-66) in (16-5) yields

$$\Delta p \geqslant \frac{h}{4\pi} \frac{2\pi p}{\frac{1}{5} \times 2h} = \tfrac{5}{4}p . \tag{A-67}$$

Thus the momentum uncertainty corresponding to an optical resolution of one-fifth of an orbital diameter is as great as the original momentum of the electron. This means that there is no way of knowing whether this "photographed" electron is still a part of the classical atom or has been knocked out of its orbit.

Answers to some questions

Chapter 2

1. It is a matter of definition. Right- or left-handedness is *defined* as that property which is inverted by a mirror. Up or down, on the other hand, is defined with respect to the direction of a gravitational field. Right or left is specified in terms of a "forward" direction (like that of a right-hand screw) which is reversed by a mirror. But the only way to change up into down is to reverse the force of gravity.

3. There is the problem of providing instructions for decoding the radio message.

5. You will have to carry out an experiment involving the weak interaction, like cobalt-60 decay.

7. Again this will require a weak interaction experiment.

Chapter 3

1. Decreased. Yes, when you consider the rest of the universe.

3. The combination of charge conjugation *(C)* and time reversal *(T)*.

4. He would have to "remember" those things which comprised his "past" before the reversal; but now we would call it the "future." However, the normal definition of memory becomes rather questionable in the context of such a reversal.

Chapter 7

4. No.   5. No.

Chapter 8

1. At 3:30 the Platform Outposts send a report by radio (or light signals) to the effect that the Rocket alert is *not* over. This is the result of the fact that clocks on the Rocket run slow as observed in the Platform frame of reference, and the countdown process is itself a clock.

   The Platform Strategists, upon receiving this message, launch their antimissile missiles. The symmetric situation, of course, occurs for the Rocket frame, which observes the Platform clocks to be running slow. The two sets of missiles intercept each other, and the resulting flashes signal both sides to release their strategic missiles.

2. None of the above. It depends on the reference frame.

3. The message was transmitted at 3:30 by the Platform Outposts from the vicinity of the Rocket. Since light traveled at *twice* the speed of the Rocket with respect to the Platform, it must have taken half as long to return as the Rocket took to reach this position. Thus the initial launching occurred at 3:45 in the Platform frame. On the other hand, the Rocket People's launching, which likewise occurred at 3:45 in *their* frame of reference, was necessarily delayed in the platform frame because, as observed by the Platform People, the Rocket clocks ran slow; hence a 45-min interval on the Rocket clock appeared as a longer interval in the Platform frame. According to equation (8–2) it was in fact

$$T = \frac{45 \text{ min}}{\sqrt{1 - (\frac{1}{2})^2}} = 52 \text{ min.}$$

   Therefore in the Platform frame the Rocket People's missiles were launched at 3:52, 7 min after the Platform's missiles, and the Platform People did indeed get the "jump" on the Rocket People, as they hoped they might.

4. Of course in the Rocket frame it was all just the other way around. Now it was the Rocket People's missiles which were launched at 3:45, and the Platform's missiles at 3:52. Thus the Rocket People too were able to get the "jump" on the other fellows.

5. In the Platform frame the Rocket's antimissile missiles were launched at 3:52. Information about this launching traveling to the Platform at the speed of light would have taken just half of the 52 min which the

Rocket had required to reach this point in space. Hence the earliest
the Platform People could have learned of the Rocket People's launching was 26 min later, at 4:18. Since the Platform People made their
initial launching at 3:45, obviously they could not have been in-
fluenced by what the Rocket People had done.

Chapter 9

1.   900,000,000 gallons.

2.   Raising numbers to such high powers as 10 and 30 is most readily
     accomplished with a loglog slide rule, or by the use of tables.

     $(0.98)^{10} = .817,$ $(0.98)^{30} = .545$

Chapter 13

1.   No. Even though the pulses have canceled each other momentarily,
     the elements of the rope where they were last seen are still in motion,
     and therefore have kinetic energy, even though their displacement is
     zero.

2.   The "memory" takes the form of the instantaneous velocity of each
     element of the rope.

Chapter 14

1.   If the atom is a sphere of radius $r = 10^{-10}$ m, the area of the surface
     presented to the light source is πr^2, namely, the cross-sectional area
     of the sphere. (This is the equivalent flat area which will absorb
     radiation at the same rate as the sphere.) To determine the amount of
     light intercepted by the atom, consider the atom to be mounted on the
     inside of a large imaginary sphere of radius $R = 1$ m. Then the fraction
     of the emitted energy absorbed by the atom is the ratio of its cross-
     sectional area to the inside area $4\pi R^2$ of the large sphere. In other
     words,

     $\pi r^2/4\pi R^2$

     is the fraction of one joule per second intercepted by the atom; the
     rest of the radiation escapes and is lost. The time it takes the atom to
     absorb 10^{-19} joule of energy is 10^{-19} joule divided by the number of

joules per second it absorbs,

$$\text{Time} = \frac{10^{-19} \text{ joule}}{(\pi r^2/4\pi R^2) \times 1 \text{ joule/sec}} = \frac{10^{-19}}{(10^{-10})^2/[4 \times (1)^2] \times 1}$$

$$= 40 \text{ sec.}$$

2. Solve equations (14–3) and (14–6) simultaneously, using equal masses $m_1 = m_2 = m$, and with one of the masses having zero velocity before the collision.

Chapter 15

1. $\lambda = 6.6 \times 10^{-37}$ cm. The distance between tunnels would have to be not too much larger than this.

2. Such a small value of Planck's constant would make their world more "classical"; one would thus expect it to be more difficult for them to discover quantum mechanics, because of the reduced de Broglie wavelength for particles. And the lines seen in the atomic spectra would be so close together that their discrete quantum nature would be unobservable to people like ourselves. But this is all based on the assumption that the change of Planck's constant affects only the events being observed and not the observers or the apparatus. On the other hand, whereas a particle wavelength is proportional to h, the radius of a Bohr orbit is proportional to the *square* of h, as shown in equation (A–58) of the Appendix. Since atoms would be so much smaller, and hence packed so much more tightly, the closer atomic lattice spacing would give them much finer diffraction gratings for their Davisson-Germer experiment. If we allow for such effects, the eventual net result is difficult to assess; it is especially problematical whether living creatures so much reduced in size are even possible. The question had therefore best be answered without this complicating dimensional change in the observers or the measurement apparatus.

3. If the dimensions of objects were somehow to remain unaffected, the bowling alleys would be a nightmare and the railroad engineers had better learn quantum mechanics.

Chapter 16

1. In order to return to earth the space ship must attain the same orbital velocity as the earth itself. The astronaut must likewise pinpoint the position of the earth in this orbit. But the uncertainty principle imposes a minimum uncertainty product $\Delta x \, \Delta p$ in position and momentum

which is equal to $h/4\pi$. For the hydrogen atom the product of the electron momentum and the characteristic dimension (the radius) of the orbit is, according to equation (15–3),

$$mvr = \frac{h}{2\pi} = \frac{6.6 \times 10^{-34} \text{ joule-sec}}{2\pi} \approx 10^{-34} \text{ joule-sec.}$$

Since this position-momentum product is comparable in magnitude to the position-momentum uncertainty $h/4\pi$, the astronaut would have a great deal of difficulty in effecting a "landing" on the electron. But in the case of the earth's orbit the characteristic product is much larger; it is in fact

$$mvr_{(\text{earth})} = 6 \times 10^{24} \text{ kgm} \times 3 \times 10^4 \text{ m/sec} \times 10^{11} \text{ m}$$

$$= 1.8 \times 10^{40} \text{ joule-sec.}$$

This is larger than the uncertainty product

$$h/4\pi \approx 0.5 \times 10^{-34} \text{ joule-sec.}$$

by a factor of something like 4×10^{74}, and it is for this reason that the uncertainty principle does not influence the space program. Even if we were transplanted to Brobdiplanck, where h is 10^{40} times greater than it is now, the uncertainty product would still be smaller than $mvr_{(\text{earth})}$ by a factor of more than 10^{34}. Thus the astronaut would have no trouble getting back to earth even in this eventuality.

2.   a) Nothing can be known about its position; the uncertainty Δx *is* infinite.   b) Nothing.   Now Δp is infinite.

Reading list

Bridgman, P. W., *The Logic of Modern Physics*, The Macmillan Company, 1949, Copyright 1927.

Best discussion I have seen of the operational definition and how it relates to the developments of physics. It is most unfortunate this book is now out of print.

Einstein, A., and others, *The Principle of Relativity*, Dover Publications, Inc., 1927. (Paperback)

For reading by physicists only, but worth looking at because it contains the original papers. Simplicity and brevity of Einstein's proof of mass-energy equivalence is dramatic, since it is completely contained on pp. 69–71.

Ford, Kenneth W., *The World of Elementary Particles*, Blaisdell Publishing Co., 1963. (Paperback)

Readable and well written treatment not only of the elementary particles but related developments in modern physics.

Ford, Kenneth W., *Basic Physics*, Blaisdell Publishing Co., 1968.

Includes the previous book and is expanded into a complete textbook on both classical and modern physics for nonscientists.

Fromm, Erich, *Escape from Freedom*, Avon Publishing Co., 1941. (Paperback)

Psychological problems posed for people living in these times by the sudden rush of intellectual and physical freedom.

Gamow, George, *Mr. Tompkins in Wonderland,* Cambridge University Press, 1957.

Adventures in a world where the speed of light is small enough and Planck's constant large enough to make the effects of relativity and quantum mechanics, respectively, unmistakable in everyday living. Anything by Gamow is worth reading.

Guillemin, Victor, *The Story of Quantum Mechanics,* Charles Scribner's Sons, 1968.

The quantum-mechanical story is summarized and still leaves room for brief accounts of some of the more recent developments. A final section discusses philosophical questions such as determinism, causality, free will.

Heisenberg, Werner, *Physics and Philosophy,* Harper & Row, 1958.

Philosophical interpretation of modern physical discoveries, by one of the men who made them happen. Required reading for anyone interested in philosophy, but becomes especially meaningful if you have read at least one book on modern physics.

McLuhan, Marshall, *The Medium is the Message,* Random House, 1968.

The radical changes in human thought and behavior effected by the electronic age, particularly in a generation reared on television. Not everyone may be as willing as McLuhan to accept it all as the wave of the future, but he shocks us into seeing what has been wrought.

Sakharov, A. D., *Progress, Coexistence, and Intellectual Freedom,* translated by the New York Times, W. W. Norton & Co., Inc., 1968.

A leading Soviet physicist discusses nuclear war, freedom, and the establishment. Has not been published in the USSR, but typewritten manuscript circulated extensively in the scientific community. The author is believed to be unofficial spokesman for many Soviet scientists, particularly the younger ones.

Sartre, Jean-Paul, *Existentialism and Human Emotions,* French & European Publishers, New York, 1968. Philosophical Library, Inc., 1957.

Defines such things as existentialism, humanism, responsibility, and freedom, by making use of the operational criterion, which Sartre calls "action."

Schrödinger, Erwin, *What is Life?* Cambridge University Press, 1963.

Co-founder of quantum mechanics speculates on the nature of life and free will in the context of the new physical discoveries.

Snow, C. P., *The Two Cultures and the Scientific Revolution,* Cambridge University Press, 1961.

This very brief and controversial work advances the thesis that solution of the problems of our times requires breaking the existing educational dichotomy between the scientific and literary cultures, whose representatives are abysmally ignorant of each other's work and presently cannot even communicate.

Spender, Stephen, *The Struggle of the Modern,* University of California Press, 1963. (Paperback, 1965)

A writer confronts the world left by the scientific revolution. In this discussion of current trends in the arts, he has little use for the benefits of technology proclaimed by C. P. Snow, whom he evidently suspects of being somewhat patronizing. Suggests that "poetic imagination" may actually be harmed by too much intellectual knowledge, and that it is the "reason and logic" of the scientists which may destroy civilization.

Vonnegut, Kurt, *Cat's Cradle,* Holt, Rinehart & Winston, 1963. (Also available as paperback)

In this powerful satiric whimsy a physicist discovers means for destroying all life on earth, and delivers the device to his children to play with. Vonnegut is my favorite antiphysicist.